# FLAME SPECTROMETRY IN ENVIRONMENTAL CHEMICAL ANALYSIS: A PRACTICAL GUIDE

# RSC Analytical Spectroscopy Monographs

Series Editor: Neil Barnett, *Deakin University, Victoria, Australia.*

Advisory Panel: F. Adams, *Universitaire Instelling Antwerp, Wirijk, Belgium*; R. Browner, *Georgia Institute of Technology, Georgia, USA*; J. Callis, *Washington University, Washington, USA*; J. Chalmers, *ICI Materials, Middlesborough, UK*; J. Monaghan, *ICI Chemicals and Polymers Ltd., Runcorn, UK*; A. Sanz Medel, *Universidad de Oviedo, Spain*; R. Snook, *UMIST, Manchester, UK.*

This series aims to provide a tutorial approach to the use of spectrometric and spectroscopic measurement techniques in analytical science, providing guidance and advice to individuals on a day-to-day basis during the course of their work with the emphasis on important practical aspects of the subject.

Flame Spectrometry in Environmental Chemical Analysis: A Practical Guide by Malcolm S. Cresser, *Department of Plant and Soil Science, University of Aberdeen, Aberdeen, UK*

*Forthcoming titles*

Chemometrics in Analytical Spectroscopy by Mike J. Adams, *School of Applied Sciences, University of Wolverhampton, Wolverhampton, UK.*

Induced Plasma Sources for Mass Spectrometry by Hywel Evans, *Department of Environmental Sciences, University of Plymouth, Plymouth, UK*; Jeffrey Giglio, Theresa Castillano, and Joseph Caruso, *McMiken College of Arts and Science, University of Cincinnati, Cincinnati, USA.*

*How to obtain future titles on publication*

A standing order plan is available for this series. A standing order will bring delivery of each new volume immediately upon publication. For further information, please write to:

Turpin Distribution Services Ltd.
Blackhorse Road
Letchworth
Herts. SG6 1HN

Telephone: Letchworth (0462) 672555

RSC
ANALYTICAL
SPECTROSCOPY
MONOGRAPHS

# Flame Spectrometry in Environmental Chemical Analysis: A Practical Guide

**Malcolm S. Cresser**
*Department of Plant and Soil Science, University of Aberdeen,
Aberdeen, UK*

THE ROYAL
SOCIETY OF
CHEMISTRY

A catalogue record for this book is available from the British Library.

ISBN 0–85186–734–0

Published by The Royal Society of Chemistry,
Thomas Graham House, The Science Park, Cambridge CB4 4WF

Typeset by Vision Typesetting, Manchester
Printed and bound Bookcraft (Bath) Ltd.

# *Preface*

Most books on analytical atomic spectroscopy written over the past few years have tended to be a bit like Christmas celebrations, with the bulk of the content not really new, but moments of inspiration or novelty occasionally shining through. They nevertheless provide a useful service if well written, in so far as they provide an up-to-date account for newcomers to a field or novices wishing to know a little more.

Most such books tend to follow the same formula, with a dabble in relevant history, a thick wad of impressively complex-looking theory for those who are easily impressed by such material, an in depth overview of instrumentation, and finally a chunky section on how to analyse particular samples or to determine particular elements.

Many of the environmental scientists who use flame spectroscopy quite extensively have neither the time nor the desire to be able to learn enough to answer Trivial Pursuit type questions on the theory or history of atomic spectroscopy. They wish instead to be able to obtain, rapidly, safely, inexpensively, and with minimal mental effort, analytical results of sufficient accuracy and precision to meet a specific purpose. It is for this group that this book is predominantly written, as a simple practical guide. It is based upon more than 20 years of experience of helping such people, primarily in a University context, take their first faltering steps in atomic spectrometry. Most have emerged with sound data, some useful knowledge, and increased confidence and job skills. Over this period, although I have made a few mistakes, I have also picked up some useful tips and ideas. It is the latter which I hope will be passed on to the attentive reader.

Malcolm S. Cresser
Aberdeen
January 1994

# Contents

**Chapter 8    How Do I Know I'm Getting the Right Answer?    95**

**Chapter 9    Safety in Flame Spectrometry                         99**

**Subject Index                                                      103**

CHAPTER 1

# *What is Flame Spectrometric Analysis?*

## 1 The Nature of Light and Heat

Light and heat are both forms of energy, and as such can interact with matter to produce physical changes in the matter. Sometimes the nature of such interactions is obvious, or even dramatic, to onlookers; for example, many ceramic materials may shatter violently if heated rapidly by a flame. Often the effects are more subtle, such as when light from a burglar's torch activates an alarm or when light reflected from objects produces changes within the eye which the brain ultimately interprets as colour and brightness.

Frequently energy is transformed from one form to another. Use of a telephone involves a large number of such transformations, from conversion of chemical potential energy stored in the body, via movement, to sound waves, which in turn pass kinetic energy (energy associated with movement) on via the mouthpiece to produce electrical signals which pass along a wire, and so on. Although they are complex, because of their familiarity we tend to take such energy transfers and transformations for granted.

## 2 Interaction of Light and Heat with Atoms and Molecules

When visible light energy interacts with coloured molecules, or in other words when it is absorbed, this physical change takes the form of redistribution of some of the electrons of the molecule to different electronic orbitals. In this way the molecule temporarily stores the absorbed energy. We say it is 'electronically excited'. Atoms may become 'electronically excited' in a similar fashion, although this involves the absorption of ultraviolet (UV) light rather than visible light in most instances. Similarly, if molecules or atoms are subjected to high temperature environments, some of the heat energy may likewise be converted to electronic excitation energy.

The electronically excited molecules and atoms generally are not very stable, and soon lose their extra energy. They mostly do this by passing the energy on to other molecules by interaction with vibrating and rotating atoms bound within those molecules. For molecules in solution, for example, the energy transfer is

1

mainly to solvent molecules, but such collisional deactivation will also occur in systems in the gaseous phase. Some molecules or atoms may be able to get rid of some of their extra energy by the emission of light energy, known as luminescence. The latter takes two forms, depending upon whether metastable intermediate electronically excited species are produced. If they are, the emission of light, then known as 'phosphorescence', is delayed by anything up to a few seconds. 'Fluorescence' emission, in the absence of metastable intermediate states, is much more rapid, generally occurring within a small fraction of a microsecond.

# 3   A Simple Quantitative Description of Light

In many situations, the behaviour of light can be best explained if the light is treated as an energetic wave. The wavelength of the light, $\lambda$, is a measure of its energy. Expressed mathematically, the energy $(E)$ is proportional to the speed of light $(c)$ divided by the wavelength $(\lambda)$, or $E = hc/\lambda$, where $h$ is a constant known as Planck's constant. Light at the red end of the spectrum has much less energy available than light at the violet end of the spectrum. Thus less energy is required to thermally excite atoms to electronically excited states which are likely to lead to the emission of red light than is required to excite them to states likely to lead to the emission of violet or UV light. The significance of this will become clearer in Chapter 2, when the instrumental requirements of flame emission spectrometry (FES) are discussed.

Visible light represents only a very small part of a much broader spectrum of energies. At wavelengths below the violet (*i.e.* with higher energies) come the ultraviolet (UV), and then $X$-rays and gamma rays. At wavelengths longer than red light come infrared (IR), and then radiowaves and microwaves. Only the UV–visible region is of interest in the present context, because this is the light which causes the physical changes in atoms or molecules which are exploited in analytical flame spectrometry.

# 4   The Absorption of Light for Quantitative Measurement

For coloured solutions, it is intuitively obvious that there is some sort of relationship between concentration of a soluble coloured species and the colour of the solution; the darker your tea or coffee, the stronger it is. Such solutions appear coloured because they absorb visible light. If they did not, they would be transparent. If a solution looks green, it must be transparent to green light. It may also transmit some blue and yellow light. It must, however, be selectively absorbing orange, red, and purple light. Measurement of the amount of red, orange, or purple light absorbed can provide a means of measuring the concentration of the absorbing species.

The human eye is not normally sensitive to UV light, and therefore notices nothing if UV light only is being absorbed. Atoms of many elements absorb light mainly or only in the UV region of the spectrum, and this region of the spectrum is widely exploited in atomic absorption spectrometry (AAS). At high concentra-

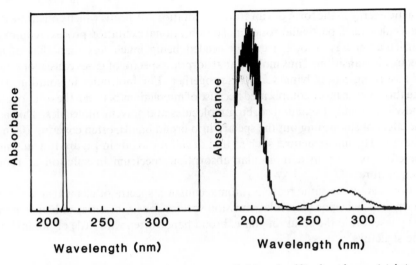

**Figure 1** *Absorption spectra of atomic zinc (left) and sulfur dioxide gas (right)*

tions especially, absorption of visible light may become significant for some elements. For example, the spectrum of our sun shows a number of dark absorption bands, where the continuum emitted from the high temperature solar surface is selectively absorbed by free atoms of elements such as sodium present in the solar atmosphere. These dark lines, the Fraunhoffer lines, are perhaps the oldest and best-known example of atomic absorption.

## 5 The Difference between Atomic and Molecular Absorption

It is appropriate at this point to consider the difference between the absorption of light by atoms and by molecules. Figure 1 compares the absorption spectrum of an element such as zinc with that of a simple triatomic gas, sulfur dioxide. The absorption spectra show how absorbance (a term explained more fully in section 1 which indicates how strongly light is being absorbed) varies with wavelength. Any particular electronic transition in an atom or molecule requires photons with an appropriate amount of energy to bring about a transition from a lower discrete (quantized) energy state to a higher quantized energy state. If the photons do not have enough energy, in other words, if the wavelength of the light is too long, the transition cannot occur. For atoms, the transition cannot occur if the wavelength is too short, either, for there is no mechanism by which the excess energy may be absorbed. Atomic absorption spectra therefore consist of isolated, very narrow bands, or lines, with one line for each possible electronic transition. This explains why the atomic absorption bands of sodium in the sun's atmosphere are sharp lines.

In the case of molecular absorption, the excess energy may be transferred into kinetic energy in the form of vibrations or rotations of atoms which constitute the molecules. Each particular rotational or vibrational excitation process requires a particular energy also, the energy needed being much less than that of an electronic transition. Thus molecular absorption spectra of gases consist of series of absorbing bands lying very close together. For molecules in solution, the situation is even more complicated because of interactions between the rotationally and vibrationally excited absorbing molecules and solvent molecules. This has the effect of smoothing out the spectra into broad bands, often covering 100 nm or more. The fine structure, such as that of sulfur dioxide in Figure 1, would not therefore be seen in a molecular absorption spectrum in solution at room temperature.

For exactly the same reasons, atomic emission spectra of elements consist of sharp lines, whereas molecular emission spectra consist of series of bands in the gas phase, or, in the solution phase, broad bands often with little or no obvious fine structure.

## 6   Quantitative AAS—What Should We Measure?

The wavelengths at which absorption or emission spectral peaks occur are characteristics of the particular atom or molecule giving rise to the peaks, and thus may be used for qualitative identification. In quantitative instrumental methods of analytical chemistry, we try to measure some property of atoms or molecules which varies linearly with the concentration of the species of interest. What parameter should we measure if we wish to exploit atomic absorption?

Consider a number of photons ($I_0$) in a narrow, monochromatic beam passing through a cloud of ($n$) atoms: If some photons are absorbed by the atoms in the cloud, the number of transmitted photons ($I_t$) will be less than $I_0$. Thus:

$$I_t = xI_0, \text{ where } 1 > x > 0 \tag{1}$$

Consider now what happens if the concentration of atoms in the atom cloud is doubled. Suppose the probability of a photon being absorbed is independent of the number of photons, and depends only upon the number of atoms in the beam path. $I_0 - I_t$ photons will still be absorbed by the first $n$ atoms, so we need to consider how the additional $n$ atoms will interact with the $xI_0$ photons which were **not** absorbed by the first $n$ atoms. Thus for the second $n$ atoms, once again a fraction '$x$' of these $xI_0$ photons will be absorbed. Thus when '$n$' is increased to '$2n$', $I_t$ is then given by $x^2I_0$. Similarly, for $n, 2n, 3n, 4n \ldots$ atoms, $I_t$ would have the values $xI_0, x^2I_0, x^3I_0, x^4I_0 \ldots$ Thus the relationship between $I_t$, $I_0$, and the atom concentration, $c$, is of the form:

$$I_t = x^cI_0 \tag{2}$$

or: $$\log I_t = c \log x + \log I_0 \tag{3}$$

or: $$\log I_t = \log I_0 = kc \tag{4}$$

or: $$\log (I_t/I_0) \text{ is proportional to } c \tag{5}$$

Since $I_t < I_0$, $I_t/I_0 < 1$, and $\log (I_t/I_0) < 0$, if we define a parameter $A$, the absorbance, as:

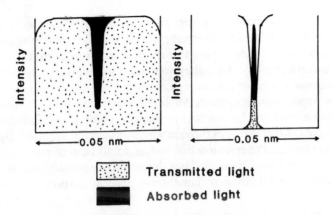

**Transmitted light**

**Absorbed light**

**Figure 2** *Schematic representation of the spectral region isolated by the monochromator when a continuum source is used (left) and when a spectral line source is used (right)*

$$A = - \log (I_t/I_0) \qquad (6)$$

the absorbance, $A$, will always be positive and proportional to concentration. Thus absorbance is the parameter which should be measured if straight line calibration plots are deemed desirable. Note that if 90% of the light is absorbed, $I_t/I_0$ is equal to 10/100 or 0.1, and $A$ is equal to 1. Similarly 99% absorption corresponds to an absorbance of 2, and 99.9% to an absorbance of 3, and so on. If you think about it, it should become apparent that precise measurement of absorbance values $> 2$ is likely to be difficult in practice.

## 7  The Sensitivity Problem in AAS

Although both the concept of absorbance and the nature of atomic absorption spectra had been understood for many decades by spectrophysicists by the early 1950s, atomic absorption had not been applied in quantitative analytical spectrometry at that time. The main limitation appeared to be the narrowness of atomic absorption lines. Monochromators could be used relatively easily (see Chapter 2) to provide a window to isolate bands of the UV–visible spectrum about 0.1 nm in width, while atomic absorption occurred over a much narrower spectral interval, typically 0.005 nm or less. Consider the situation if quite strong atomic absorbance occurred as a consequence of light passing through a cell such as a flame containing free atoms of the element of interest; there would be no change in 95% of the light passing through the monochromator 'window'. This situation is represented in Figure 2. Thus $I_t$ would be only very slightly smaller than $I_0$, the ratio $I_t/I_0$ would be close to 1, and, from equation 6, absorbance would be close to zero. Sensitivity would therefore always be very poor.

Sensitivity could, of course, be improved significantly by using a very high resolution monochromator, which could isolate regions of the spectrum of 0.005

nm or less. At that time, however, such monochromators were large and expensive, had a very low light throughput (which could result in signal stability problems), and sometimes required fairly precise temperature control to give good wavelength stability. Thus they were hardly ideal for routine use in busy and often congested laboratories.

A major breakthrough came in Australia when Alan Walsh[1,2] realized that light sources were available for many elements which emitted atomic spectral lines at the same wavelengths as those at which absorption occurred. By selecting appropriate sources, the emission line widths could be even narrower than the absorption line widths (Figure 2). Thus the sensitivity problem was solved more or less at a stroke, and the modern flame atomic absorption spectrometer was born.

## 8   The Potential Selectivity of Flame AAS

The importance of Walsh's ideas should not be underestimated. Not only had he suggested a potentially highly sensitive method of analysis which would prove eventually to be suitable for the determination of many elements in the periodic table, but at the same time he had suggested a development which, theoretically at least, should lead to virtually specific analysis. The very narrowness of the absorption lines which had hitherto held back progress in AAS suddenly became its most powerful asset. It meant that the chances of spectral overlap of the absorption line of one element with the emission line of another were extremely small. Thus atomic spectral interferences should be, and indeed are, rare in AAS.

It is little wonder, then, that analysts working in environmental laboratories were amongst the first to grasp the importance of Walsh's benchmark papers.[3,4] Hitherto the complex nature of their typical sample matrix and the nature of the elements they were interested in often resulted in the need for complex and time consuming separation and preconcentration by skilled chemists prior to the determination step. Suddenly they were apparently being offered a very sensitive and almost universally applicable technique where the only sample preparation needed seemed to be sample dissolution. Eventually problems did start to surface, as is invariably the case, but as Chapter 3 will show, few have proved insuperable.

## 9   Flame Emission Spectrometry

So far in this chapter, absorption techniques have been considered in preference to emission techniques, in spite of the much longer history of flame emission spectrometry (FES).[5] This is deliberate, and reflects the far greater relative importance of AAS as a routine technique in most modern environmental

[1] A. Walsh, *Spectrochim. Acta*, 1955, **7**, 108.
[2] B.J. Russell, J.P. Shelton, and A. Walsh, *Spectrochim. Acta*, 1957, **8**, 317.
[3] D.J. David, *Analyst (London)*, 1958, **83**, 655.
[4] J.E. Allan, *Analyst (London)*, 1958, **83**, 466.
[5] L.H.J. Lajunen, 'Spectrochemical Analysis by Atomic Absorption and Emission', The Royal Society of Chemistry, Cambridge, 1992.

analytical laboratories. Flame atomic fluorescence spectrometry (AFS) evolved a decade after flame AAS,[6] but is still of only minor importance. A similar balance will be seen in the following chapters of this book, since many FES and AFS determinations are performed utilizing atomic absorption spectrometers.

The introduction of the first flame spectroscopes early in the second half of the 19th century was as impressive an achievement in its time as that of AAS a century later. Elements were introduced into a flame in the form of a salt and the spectrum excited in the flame was examined using a prism spectrometer. Using the eye as a detector, characteristic emission wavelengths of elements could be measured, and subsequently used as a basis for qualitative analysis. In the flames available at the time, only those elements such as K, Li, Na, and Ca which were easily thermally excited (elements with low excitation potentials) and which therefore emitted visible light were studied. However the technique was successfully applied to establishing the existence of other elements falling into this category, such as Cs, In, Rb, and Tl.[5]

The potential development of FES as a quantitative technique was hampered until well into the 20th century by the use of the eye as a detector; this limited the reliability of quantification by the comparison of light intensity from samples and standard materials. The evolution of photocell detectors enhanced sensitivity and improved reliability, considerably extending the range of application of FES techniques.

The emission intensity depends upon the number of excited atoms, $N^*$, which in turn depends upon the number of unexcited or ground-state atoms, $N_0$, the statistical probabilities of excitation and emission occurring, and exponentially upon the flame temperature, $T$, and the reciprocal of the excitation potential, $E$. Thus:

$$N^* \text{ is proportional to } N_0 \exp\left(-E/kT\right) \qquad (7)$$

where $k$ is Boltzmann's constant. High emission intensity is favoured, as might be intuitively expected, by high flame temperature and when the excitation potential is low.

There is a major difference between AAS and FES in terms of spectral selectivity. In FES, all elements present in a sample and atomized in the flame are capable of being excited to a greater or lesser extent, depending upon their excitation potentials. Polyatomic species may also occur and be thermally excited. Thus, for interference-free analysis, it is essential to be able to isolate the determinant wavelength of interest from all other emitted wavelengths emanating from sample components. Some of the most intense thermally excited molecular emission comes from the flame gases themselves, rather than from the samples. Background and concomitant element emissions place great demands upon the monochromator used and upon the knowledge of the analyst. In AAS, the function of the monochromator is primarily to isolate the wavelength emitted by a single element line source which gives the best sensitivity from other lines emitted by that source, a much simpler task.

[6] I. Rubeska, V. Svoboda, and V. Sychra, 'Atomic Fluorescence Spectroscopy', Van Nostrand Reinhold, London, 1975.

# 10   Flame Atomic Fluorescence Spectrometry

Flame AFS combines features of both AAS and FES. The excitation of atoms is
by the absorption of light. When individual element spectral line sources are
used, the spectral selectivity should be as high as that in AAS, although scatter
may be more of a problem in AFS. Quantification is by comparison of the
intensity of fluorescence emitted by samples with that emitted by standards of
known concentration. At low determinant concentrations, it is necessary to
discriminate between small fluorescence emission signals and the background
light levels associated with thermally excited emission from the flame. Therefore
in AFS, as in FES, it is desirable to have low flame background emission. This is
discussed further in Chapter 2, where instrumental aspects of flame spectrometric
techniques are discussed.

There was a good reason that AFS attracted considerable attention around the
world when it was being developed, primarily by T.S. West and colleagues at
Imperial College, London and J.D. Winefordner and his group at the University
of Florida, Gainsville.[6] Whereas in AAS the signal is independent of excitation
source intensity, in AFS, the more intense the source, the more intense the
fluorescence. The reason for this should be clear from the assumptions made
when deriving equation 2. Unfortunately developments in sources have never
been adequate to allow this potential asset to be fully realized.

# 11   Common Features of FES, AAS, and AFS

Some readers may be wondering why these three techniques should be considered
together in this and subsequent chapters, when there are clearly distinct
differences between them. However, the three also have many features in
common, and to consider every facet of them separately would involve much
repetition. In all three techniques, signal depends upon the reproducible
production of ground-state atoms in a flame atomizer cell. In all three, sample
introduction is invariably via pneumatic nebulization (see Chapter 2). Many key
instrumental components are common to all three techniques. Finally, no one of
the techniques can be regarded as superior for every determination, so it is
important for the environmental analyst to know something about the strengths
and weaknesses of them all. Collective consideration allows a little knowledge to
go a long way!

CHAPTER 2

# Instrumentation for Analytical Flame Spectrometry

## 1 Introduction

In Chapter 1, section 11, the point was made that the main flame spectrometric methods of analysis have several factors in common, including the approach used to bring about atomization following sample dissolution, the method mainly used for isolation of the wavelength of interest, and the devices most frequently employed to convert light signals into electrical signals suitable for signal processing. Readout systems too are common to all three techniques. As a consequence, many AFS and FES measurements are made using AAS instrumentation. The approach followed in this chapter therefore has been to consider in detail the essential components of a typical atomic absorption spectrometer, as represented in Figure 1, and then to consider briefly where the requirements of FES and AFS differ from those of AAS. However more specialized aspects of AAS systems, such as the background correction systems used to eliminate some types of spectral interference problem or hydride generation techniques are considered at appropriate points in subsequent chapters.

## 2 The Modern Hollow Cathode Lamp

As explained in Chapter 1, section 7, unless a very high resolution monochromator, *e.g.* an echelle monochromator, is used to isolate a very narrow ($<ca.$ 0.005 nm) band of light from a continuum spectrum prior to absorbance measurement, the sensitivity will be very poor.[1,2] Although there are occasional reports of analysis by flame AAS using continuum sources such as xenon arc lamps, these are invariably from research laboratories. The vast majority of reported applications use single element line sources, and more than 99% of these applications use hollow cathode lamps.

Figure 2 shows a schematic representation of a typical modern hollow cathode lamp. The hollow cathode, which is made from the element of interest or one of its alloys, is at the centre of the lamp, which is filled with an inert gas, generally neon or argon, at low pressure. A high voltage, low current discharge is struck

---

[1] P.N. Keliher and C.C. Wohlers, *Anal. Chem.*, 1974, **46**, 682.
[2] M.S. Cresser, P.N. Keliher, and C.C. Wohlers, *Lab. Pract.*, May, 1975, 335.

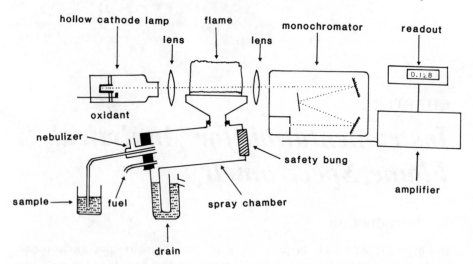

**Figure 1**   *The essential components of an atomic absorption spectrometer (not drawn to scale)*

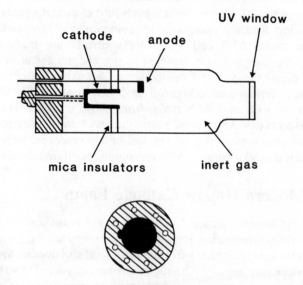

**Figure 2**   *Typical hollow cathode lamp (top) and lamp base (bottom)*

between the cathode and the anode. The latter, which is commonly made from tungsten, takes the form of a small cylinder or sometimes a small flag electrode. Sheets of insulator, often mica, confine the discharge to the central cathode region to obtain high intensity and good stability. The end window of the lamp is made from quartz or optical silica if necessary to transmit UV light. Ordinary glass starts to absorb increasingly strongly below about 320 nm.

For physical stability the lamps traditionally have an eight-pin base, which attaches to an eight-hole socket. Frequently only two of the pins serve to make any electrical connection, however, although in some modern fully automated instruments additional pins may be connected to electrical components which provide fully automated lamp identification. The lamp base itself generally has a protruding plastic lip to make sure that the lamp can be fitted only in the correct position unless excessive brute force is applied (Figure 2). So, always treat the lamps gently. They are expensive to replace!

## Lamp Alignment

When optimization is discussed in Chapter 4, it will become clear that it is very important to ensure that lamps are correctly aligned along the optical axis through the flame to the monochromator entrance slit. Many AAS instruments are designed to accommodate lamps from diverse manufacturers, and such lamps often have differing diameters. Therefore lamps normally fit inside some sort of supporting cradle. The position of the latter can invariably be adjusted in both vertical and horizontal planes. Optimal alignment generally corresponds to the maximum signal being obtained from the detector. It is tempting to use the glowing cathode image (usually red from the neon fill gas) on the entrance slit to align the lamp, but this is not advisable for final tuning. The focal point for UV light may be displaced by a few millimetres from that for red light.

## Multi-element Hollow Cathode Lamps

The vast majority of AAS determinations are conducted using single-element hollow cathode lamps, in spite of the expense involved in having one lamp for each element to be determined. This is because single element lamps tend to give superior signal-to-noise ratio, and thus marginally lower detection limits and improved precision, compared to multi-element lamps. A notable exception in the author's experience is the calcium/magnesium dual element lamp, which appears to give directly comparable performance to the corresponding single-element lamps. Multi-element lamps containing up to six or more elements are commercially available, but are not generally to be recommended if optimal performance is desired.

## 3 Alternative Line Sources

For some elements such as arsenic and selenium, which have their main atomic absorption wavelengths lying on the edge of the vacuum UV, the performance of hollow cathode lamps is often poor, the lamps displaying low intensity and poor stability. This, plus the search for more intense sources for AFS (see Chapter 1, section 10), resulted in the development of microwave-powered electrodeless discharge lamps (EDLs) as spectral line sources towards the end of the 1960s.[3-5]

---

[3] J.M. Mansfield, M.P. Bratzel, H. Norgordon, D.N. Knapp, K.E. Zacha, and J.D. Winefordner, *Spectrochim. Acta, Part B*, 1968, **23**, 389.

[4] K.E. Zacha, M.P. Bratzel, and J.D. Winefordner, *Anal. Chem.*, 1968, **40**, 1733.

[5] R.M. Dagnall and T.S. West, *Appl. Opt.*, 1968, **7**, 1287.

A few milligrams of the element of interest or one of its halides was sealed in a quartz tube a few centimetres in length and around 8 mm in diameter, containing argon at low pressure. The element spectrum was excited in an appropriate resonance cavity using microwave radiation at a frequency of around 2450 MHz and a power of around 25–50 Watts. For volatile elements such as cadmium, mercury, and zinc, the lamps were generally much more intense than the corresponding hollow cathode lamps, sometimes by two to three orders of magnitude. However, they were notoriously unstable, and required too much operator skill (a mixture of art and science!) to find favour for routine use in AAS.

More recently, radiofrequency-powered (r.f.) electrodeless discharge lamps have become commercially available for a fairly wide range of elements, including As, Bi, Cd, Cs, Ge, Hg, K, P, Pb, Rb, Sb, Se, Te, Tl, Sn, Ti, and Zn.[6] Although not as intense as some of the old microwave EDLs, r.f. EDLs exhibit far superior stability, and are often far brighter than the corresponding hollow cathode lamp. They are especially worth considering if a large number of arsenic or selenium determinations are to be performed routinely, although considerable expense is involved because the lamps require a separate r.f. power supply.

Brief mention should also be made here of high intensity (also known as 'boosted output') hollow cathode lamps.[7] In these lamps an auxiliary current of around 200–400 mA is applied to the dilute cloud of atoms sputtered outside the central zone of the normal hollow cathode. The atoms are thus excited and emit intense radiation which may be used in AAS or AFS. Once again an auxiliary power supply is required, and the lamps themselves are more complex and correspondingly more expensive. Such lamps have had a rather chequered history, finding great favour in some environmental analytical laboratories but never being widely used on any routine basis.

## 4   How to Distinguish between Lamp Light and Flame Light

When we are measuring the absorption of light by solutions, the solution itself does not generally emit any light. It is therefore easy to measure the change in $I_0$ when a light beam passes through a cell containing molecules in solution. In flame atomic absorption spectrometry, however, the device used to detect the light signal will detect both light emitted by the hollow cathode lamp and light emitted by the flame within the band of spectrum being used. Thus the detector would 'see' too much light, and absorbance would not be correctly measured. By modulating the lamp power supply so that the lamp flashes at a frequency of, say, 200 Hz, and making sure that the instrument only responds to modulated light at this frequency, it is possible to discriminate between the light from the flame (unmodulated) and the light from the hollow cathode lamp (modulated). In this way true absorbance may be measured, even when the flame itself is emitting quite intensely. Therefore the power supplies to hollow cathode lamps in atomic

[6] R.D. Beaty, 'Concepts, Instrumentation and Techniques in Atomic Absorption Spectrophotometry', Perkin Elmer, Washington DC, 1988.
[7] J.V. Sullivan and A. Walsh, *Spectrochim. Acta*, 1965, **21**, 721.

absorption spectrometers must always be modulated. The required lamp signal is isolated by synchronous demodulation.

# 5   The Flame as an Atom Cell

Because atomic absorption involves the absorption of light by free atoms, some method is necessary for conversion of the determinant species in a sample into atoms. This is most commonly done by dissolving the sample, and spraying the resulting solution into a flame which is hot enough to convert the determinant to atoms.

The flame must dry, vaporize, and atomize the sample in a reproducible manner with respect to both space and time. Unlike titrimetric and gravimetric analysis, atomic absorption spectrometry is a secondary analytical technique. Concentrations are determined by comparing the absorbance values obtained for samples with those obtained for standards of known determinant concentrations. It is very important, therefore, that samples and standards are always atomized with the same efficiency to produce a cloud of atomic vapour of highly reproducible geometry. If samples and standards behave differently, errors will result.

## The Air–Acetylene Flame

In the earliest days of flame AAS, air–propane and air–butane flames were often used to atomize samples, largely because they had a reputation for being simple and safe in operation. However it was soon found that such flames were not satisfactory for breaking up many thermally stable chemical compounds into the free atoms required to obtain atomic absorption. If samples and standards are not atomized to the same extent, erroneous results are obtained. Nowadays the most commonly used flame is the air–acetylene flame. This flame is safe and relatively inexpensive to use, and sufficiently hot at *ca.* 2200 °C to atomize molecules of many common elements. However it is not sufficiently hot to break the element–oxygen bonds of some elements, the so-called refractory oxide-forming elements. These include, for example, aluminium and silicon. Such determinants require a hotter flame. Also atomization efficiency of some elements may be influenced by matrix elements and ions. For example, phosphate or aluminium depress the atomic absorption signals of calcium in an air–acetylene flame. Thus there is a need for a safe, inexpensive and reliable higher temperature flame in AAS.

## The Problem with Some Hotter Flames

To produce a stable flame on a burner head requires a gas mixture in which the upward flow velocity just exceeds the downward burning velocity. If this situation is reversed, the flame may burn back through the burner slot or holes, resulting in a potentially dangerous explosion, a process known as a flash-back. Pre-mixed oxygen–acetylene flames are substantially hotter than air–acetylene flames, but they are never used routinely because the burning velocity is too great, and the risk of explosive flash-back is too high. If a flashback does occur with this gas mix, the explosion is very violent.

On the other hand, if the flow velocity exceeds the burning velocity by too great a margin, the flame 'lifts off' from the burner head. Many readers will have experienced this phenomenon at some stage in their career when trying to light a bunsen burner with the air hole fully open. The flame takes the form of an unstable fire ball a few centimetres above the burner port for a few seconds, and then often is extinguished. At a given total flow of fuel and oxidant, the flow velocity is regulated by the dimensions of the burner slot; the narrower and/or shorter the slot, the faster the flow velocity. To burn pre-mixed oxy–acetylene or oxy–hydrogen mixes, which have very high burning velocities, requires extremely narrow burner slots; so narrow, in fact, that the slots would be very prone to clogging for many typical environmental sample matrices. This is another reason for the non-use of oxygen-supported flames.

## The Nitrous Oxide–Acetylene Flame

For some years it was thought that a safe flame that was appreciably hotter than the air–acetylene flame would not be found, until John Willis[8] suggested the use of the pre-mixed nitrous oxide–acetylene flame (sometimes also known as the dinitrogen oxide–acetylene flame). This flame could provide a temperature of around 3000 °C and in addition, in the fuel-rich flame, an atmosphere which was chemically very reducing, and excellent for breaking refractory metal oxide bonds. Its burning velocity was greater than that of the air–acetylene flame, so that a smaller burner slot was necessary, but much less than that of the oxygen–acetylene flame. Nevertheless, in the early days of its use, occasionally very violent flash-backs occurred.

## Safe Use of Flames

In modern atomic absorption spectrometers, the spray chamber which contains the fuel–oxidant gas mixture is fitted with a safety bung or blow-out membrane. If the mixture inside the spray chamber is accidentally ignited via a flashback, the rapid pressure build up blows out the bung or ruptures the membrane, immediately releasing the pressure and minimizing the risk of damage to the mixing chamber and other instrumental components. The drain which takes away surplus solution also functions in this way to some extent. After a flash-back, it is imperative to replace the bung or membrane, and to refill the drain before attempting to relight the flame.

In many modern instruments, especially the more expensive ones, a flash-back causes a switch of some sort to trip, immediately shutting off the fuel supply to minimize the risk of fire. The switch must be reset before the flame can be relit. More sophisticated instruments incorporate additional sensors related to safety. These include devices to detect the presence of the correct burner head, gas

---

[8] J.B. Willis, *Nature* (London), 1965, **207**, 715.

pressure sensors in the fuel and oxidant lines, and even flame detectors which shut off the fuel if the flame does not appear to be alight.

It is important to be sure that all fuel and oxidant lines and their associated connectors within the instrument are in good condition, because virtually no instrument detects automatically slow fuel leaks which could cause a build up of an explosive gas mixture within the instrument casing. Obviously piping and connectors external to the instrument must also be in sound condition.

Acetylene should not be used routinely at pressures above 10 p.s.i., because detonation is possible under this condition. It should not be allowed to come into contact with copper piping or fittings, because of the risk of formation of explosive copper acetylide. Acetylene cylinders contain the gas dissolved in acetone on a porous ceramic support and must be stored upright to avoid the possibility of liquid acetylene getting into the fuel lines. If ever it does, the instrument fuel piping and connectors must be checked very carefully for leaks, if necessary after getting advice from the manufacturer of the instrument. When the operating pressure of the cylinder drops to about 80 p.s.i., it should be replaced to prevent the passage of excessive acetone vapour into the flame. The presence of too much acetone could result in greater incident or extent of interferences, as discussed in Chapter 3.

In the early days of flame AAS, the use of nitrous oxide also occasionally resulted in problems for the analyst. One was that the fumes were appreciably more toxic, so an efficient extraction hood over the flame became much more important. The nitrous oxide–acetylene flame was both hotter and taller than the air–acetylene flame, and the unwary occasionally managed to underestimate the temperature of the exhaust gases at the fan, resulting in distorted fan blades. Plastic fan blades are best avoided! It is important to use an appropriate extraction rate. The author knows of one small company which installed such a powerful extractor that the laboratory doors flew open whenever the fan was switched on. On one occasion, even with the laboratory doors open, a passing sparrow was sucked in through the window which also needed to be kept open.

A second problem with nitrous oxide was its property of cooling dramatically when allowed to expand very rapidly on going from high pressure to low pressure. This resulted frequently in ice formation on the cylinder head, and poor gas flow stability. To avoid the consequential loss in precision, cylinder heads were often warmed, or a ballast tank at an intermediate pressure could be used as a stabilizer.[9] Most modern AAS instruments employ quite high oxidant pressures and flow rates in the interests of safety, in spite of the greater cost, and this problem is less common than it used to be.

## Some Observations on Burner Heads

When flame AAS was first introduced, it appears that little in-depth thought went into the design of burner heads. The manufacturers opted for long-path burner heads (100–120 mm), because they wanted the technique to be sensitive

---

[9] M.S. Cresser and G. Wilson, *Lab. Pract.*, Feb., 1973, 117.

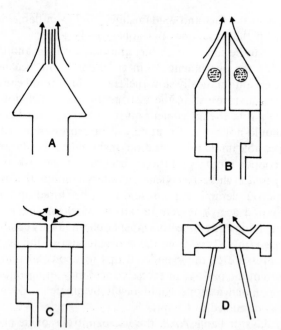

**Figure 3**   *Cross sections of: (A) Boling burner; (B) water-cooled burner head with triangular cross section; (C) flat-topped burner head; and (D) typical modern AA spectrometer burner head. Arrows denote air entrainment patterns*

and knew that longer cells gave bigger signals in solution spectrophotometry. In flame AAS the situation was rather different, because the use of a longer flame didn't put more sample into the optical path. Instead, increasing the flame cross-sectional area increased the residence time of atoms in the hollow cathode lamp beam. The burner heads were designed primarily to be mechanically robust and safe. Damaged or scared customers tend not to shop again with the same manufacturer! For ease of construction, most burner heads were flat. Flame AAS users soon learn that burner heads, and especially those that are flat-topped, rapidly get too hot to touch. Even ten minutes after the flame has been extinguished after a period of extended use, it is still possible to get a painful burn by touching the burner.

In 1966, Boling[10] published a novel burner head design with three slots rather than the normal single slot (Figure 3A). The idea was to produce, in effect, a flame-shielded flame which would have a high central flame temperature. The design was apparently successful, in so far as sensitivity for elements such as chromium, which are not readily atomized, was improved. Coincidentally, the Boling burner had an unconventional cross section. A few years later, the author was experimenting with water cooling of burner heads as a way of reducing clogging problems. However, with cooling water passing along either side of the slot in the head shown in Figure 3B, the burner head was so cold that

---

[10]  E.A. Boling, *Spectrochim. Acta*, 1966, **22**, 425.

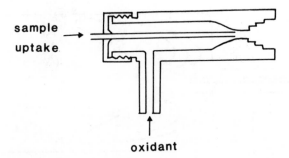

**Figure 4**  *A typical pneumatic nebulizer of the type used in flame spectrometry*

condensation in the slot soon extinguished the flame. Even with the cooling water disconnected, the burner head remained cool, so that it was possible to touch the side of the head while the flame was alight without any risk of a burn.[11]

The reason appears to be the much smoother pattern of air entrainment when the burner head has a triangular cross section (Figure 3B) compared to the very turbulent pattern and associated heating effect when a flat-topped head is used (Figure 3C). The head shown in Figure 3B gave similar sensitivity enhancements and reduced chemical interferences to the Boling burner, suggesting that the benefits of the latter were fortuitously attributable to its cross section rather than the triple slot *per se*. Most modern AAS instruments use a head somewhere between a flat-topped head and a full triangular cross section, as shown in Figure 3D.

# 6  Sample Introduction

Although we have now considered in some detail the use of flames as atomizers in analytical flame spectrometry, we have not yet considered how the sample is introduced into the flame. This is invariably achieved using a pneumatic nebulizer, which functions both as a pump for the sample solution and to break the sample up to a fine aerosol.[12] The latter is then intimately mixed with the fuel and oxidant, and transported by the latter through some sort of spray chamber to the burner head. Figure 4 illustrates a design of a typical pneumatic nebulizer. The oxidant doubles as the nebulizing gas and issues at high velocity from a narrow jet which concentrically surrounds a central capillary through which the sample solution is sucked (aspirated). Many other nebulizer designs have been suggested over recent years,[13] and the complex underlying operational theory has been extensively investigated,[14,15] but at the time of writing simple concentric pneumatic nebulizers remain central to the successful operation of the vast majority of flame-based analytical atomic spectrometers.

[11] M.S. Cresser, *J. Anal. At. Spectrom.*, 1993, **8**, 269.
[12] I. Lopez Garcia, C. O'Grady, and M. Cresser, *J. Anal. At. Spectrom.*, 1987, **2**, 221.
[13] M.S. Cresser, in 'Sample Introduction in Atomic Spectroscopy', ed. J. Sneddon, Elsevier, Amsterdam, 1990, p. 13.
[14] B.L. Sharp, *J. Anal. At. Spectrom.*, 1988, **3**, 613.
[15] B.L. Sharp, *J. Anal. At. Spectrom.*, 1988, **3**, 939.

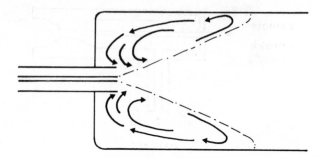

**Figure 5** *Turbulent air and aerosol re-entrainment patterns in a spray chamber*

## Transport Efficiency

Once droplets of aerosol reach the flame, the solvent, most commonly water, evaporates very rapidly, leaving minute solid particles. These must then be vaporized, and finally atomized. However the bulk of the droplets never reach the flame at all, but instead are lost by deposition onto the walls of the spray chamber. It might be expected that most of the loss would occur on the end wall of the spray chamber opposite the nebulizer, but the exact reverse occurs in practice. This has been demonstrated by spraying coloured dyes into chambers lined with absorbent paper and looking at the dye distribution.[16] Much of the loss is at the nebulizer end of the spray chamber, as a consequence of turbulence and recirculation of aerosol into the expanding cone of spray. Direct evidence for this comes from the use of smoke tracers with progressively truncated spray chambers.[16] Figure 5 shows the air recirculation pattern which occurs once the chamber exceeds a certain length. As a consequence of this loss, at normal aspiration rates the transport efficiency (the ratio of amount of determinant reaching the flame per second to the amount aspirated per second, expressed as a percentage) is generally only around 4–8%.

Over the years many analytical spectroscopists have attempted to improve upon this situation, but the only reliable way to improve transport efficiency with pneumatic nebulizers is apparently to restrict the aspiration rate.[17,18] Reduced aspiration rate means that the nebulizer energy is distributed to less aerosol per unit time, resulting in a finer droplet size distribution; finer droplets (*e.g.* <2 μm in diameter) are more likely to be transported through the spray chamber. Alternatively, the determinant may be introduced to the flame in gaseous form, or in a small cup. Such approaches are discussed in Chapter 6. However often the approach taken is to use electrothermal atomization rather than a flame,[6,19] but this is outwith the scope of the present small volume.

[16] C.E. O'Grady, I.L. Marr, and M.S. Cresser, *Analyst (London)*, 1985, **110**, 729.
[17] M.S. Cresser and R.F. Browner, *Anal. Chim. Acta*, 1980, **113**, 33.
[18] M.S. Cresser, *Anal. Proc.*, 1985, **22**, 65.
[19] L.H.J. Lajunen, 'Spectrochemical Analysis by Atomic Absorption and Emission', The Royal Society of Chemistry, Cambridge, 1992.

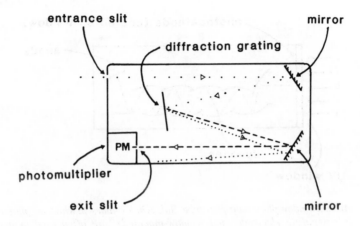

**entrance slit**         **mirror**

**diffraction grating**

PM

**photomultiplier**

**exit slit**         **mirror**

**Figure 6**    *Optical layout of a typical monochromator*

## 7 The Need for a Monochromator

A hollow cathode lamp emits an intense line spectrum of the cathode element, of any other element present in the cathode, and of the filler gas (neon or argon). It is therefore necessary to be able to isolate the lines of the determinant element from any other emitted lines. If we do not, the difference between $I_t$ and $I_0$ will be greatly reduced, and the sensitivity unacceptably poor. Moreover, not all lines of the determinant element give equal sensitivity, and it is therefore also desirable to isolate the determinant line at the wavelength which gives the most useful sensitivity from all other lines. This is done with a grating monochromator. Figure 6 illustrates a typical optical layout in the monochromator of an atomic absorption spectrometer.

In the monochromator, the diffraction grating produces a spectrum in the plane of the exit slit. The exit slit serves as a window to isolate the particular line (wavelength) of interest. When the wavelength setting of the monochromator is adjusted, the grating slowly rotates and the spectrum moves sideways across the exit slit. This adjustment may be done manually, or sometimes, in more expensive automated instruments, under microprocessor control via a stepper motor.

## 8 Detection of the Light Signal

The light signal is invariably converted to an electrical signal by a photomultiplier tube. Figure 7 shows how a photomultiplier works. When photons hit the photocathode, electrons are emitted. These are accelerated towards another electrode (the first dynode), which is held at a positive potential relative to the cathode. Each electron from the photocathode causes the emission of more than one electron from the first dynode. These electrons in turn are accelerated towards the second dynode, where each in turn causes the emission of more than one electron; and so the process continues to the last dynode.

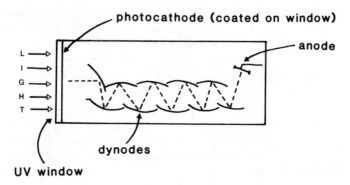

**Figure 7** *An end-window photomultiplier tube. Side-window tubes, in which the photocathode is a separate electrode (not window-mounted), are often used in analytical spectrometers*

Suppose, in a ten dynode chain, each primary electron causes the emission of five secondary electrons. After the first dynode we would have five electrons, after the second 25, after the third 125, and so on. At the tenth dynode we would have $5^{10}$ electrons. Thus the device is very sensitive. Photomultiplier tubes are available which respond well over the entire UV–visible region of the spectrum. The tubes are stable and long-lived, which is just as well, because they are expensive to replace. They require a well-stabilized high voltage power supply, which adds to the overall cost of the instrument.

It should be clear from the above simple explanation of how the devices work that their sensitivity (gain) may be regulated by altering the applied voltage. They have a wide linear response range, but do eventually become saturated. At high light levels, they often start to respond negatively to further increases in light intensity. This may cause confusion among novices in AAS when attempting to set up an instrument, so it is a good idea to have an idea of what the approximate gain setting should be on the instrument being employed.

## 9 Readout Systems

The signal from the photomultiplier is passed to a phase-sensitive, and often frequency-tuned amplifier, which isolates the component attributable to light from the hollow cathode lamp from light emitted from the flame or stray daylight, as discussed in section 4.

In the early days of analytical atomic absorption, this signal was fed directly to a meter calibrated in absorbance units, but the scale was logarithmic, and difficult to read at absorbance values $> 0.6$, because the divisions became closer and closer together. Nowadays absorbance is invariably calculated electronically, and the value read from an analog or digital scale, or a computer, or fed to a chart recorder or printer.

On inexpensive instruments, absorbance is measured for a series of standards, and then for a series of samples. The values for the standards are used to draw a calibration graph, which is then used to estimate the concentrations of determinant in the samples from their measured absorbance values. On more expensive instruments, the calibration data may be stored on a microcomputer or microprocessor, which is also used to calculate determinant concentrations directly. The calibration graph may be displayed on a monitor and/or printed, if required. It is always a good idea to take a look at the calibration graph prior to calculating results, to make sure that the standards are giving the expected results, and, for example, that the calibration graph is not showing excess curvature towards the concentration axis. Some instruments allow several readings to be taken on each sample and standard, and print out the standard deviation for the apparent concentration of each sample. This is a useful indicator of sudden drift problems, although a low standard deviation does not necessarily indicate that the result is correct. It is important always to discriminate between precision and accuracy!

## 10 Double-beam Instruments in AAS

Source stability is very important in AAS, because the instrument data processor in the instruments considered so far stores values of $I_0$ to allow continuous calculation of absorbance from continuously monitored values of $I_t$. If the source becomes less bright, the detector response will be equivalent to a positive absorbance signal. If the source becomes brighter, the source drift will be detected as negative absorbance. One way around the problem of drift in source intensity is to employ a double-beam instrument. The light beam from the hollow cathode lamp is split into two parallel beams resolved in time, one passing to the detector through the flame as usual, the other passing alongside the flame. The latter then provides a continuous record of $I_0$, the former a continuous record of $I_t$.

It should be noted that this system compensates only for drift in source intensity assuming the source line profile being used remains constant in shape, as discussed in Chapter 4. It does not compensate for fluctuations in flame background absorption. Thus it is important to re-zero instruments very frequently. Often this is done between each reading while aspirating deionized water or an appropriate solvent blank. It is also important to check for drift in signal resulting, for example, from fluctuations in aspiration rate. This is done by periodic aspiration of a standard of known concentration, typically after every 5–8 samples, and adjusting the instrument amplifier scale expansion as and when required. Thus, double-beam instruments improve precision when precision is limited by source intensity fluctuations.

## 11 Instrumental Requirements in FES

At the present time, the majority of elemental determinations conducted by FES are performed using instruments designed primarily for AAS. The only modification required is the incorporation of an amplifier capable of measuring the unmodulated emission signals from the flame, a standard feature on almost all AAS instruments.

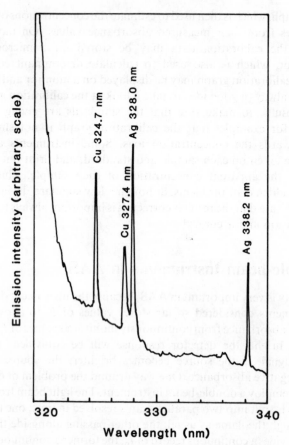

**Figure 8**  *Part of the emission spectrum from an air–acetylene flame into which a
solution containing 20 mg $l^{-1}$ of each of copper and silver was being nebulized*

The photomultiplier detects both the thermal emission from the determinant
and also any other atomic or molecular emission from either concomitant
elements present in the sample or from the flame itself. Figure 8, for example,
shows a typical section of a flame emission spectrum. While it is possible for some
determinations by FES to work at a single fixed wavelength, as in flame AAS, it is
advisable, at least initially, to scan the emission spectrum in the vicinity of the
wavelength of interest to confirm the absence of spectral interferences. In any
event, regular re-zeroing and aspiration of an appropriate standard to check for
signal drift is essential.

In the early days of AAS, most manufacturers supplied special burner heads to
be used for FES determinations. These were usually cylindrical with a circular or
square array of small holes in the flat top surface. They were based on the
traditional burner head designs of flame emission spectrometers. However often
superior results are obtained with long path burner heads.[20]

[20] M.S. Cresser, P.N. Keliher, and G.F. Kirkbright, *Sel. Ann. Rev. Anal. Sci.*, 1973, **3**, 139.

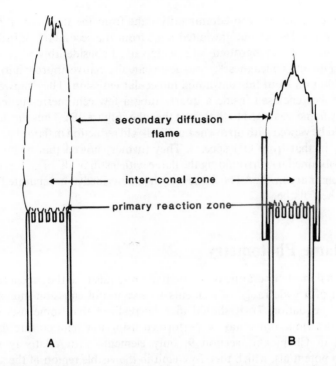

**Figure 9** *The three key zones of ordinary ( A ) and separated ( B ) air–acetylene flames. The inter-conal zone is the most important zone in analytical flame spectrometry*

Just as the nitrous oxide–acetylene flame represented a major advance in AAS, its introduction was also very important in FES. It was not only capable of atomizing many refractory oxide-forming element, but also it was a manageable and relatively safe high temperature flame well suited to determinations by FES. Some excellent detection limits were reported using this flame.[21] For elements with their principal emission wavelength(s) at above around 330 nm, FES using a nitrous oxide–acetylene flame often yielded significantly better detection limits than flame AAS. The improvement became even more pronounced at much longer wavelengths, because of the low excitation potentials associated with those wavelengths.

## Separated Flames

Figure 9A shows the structure of a typical pre-mixed air–acetylene flame. The zone of interest to us is the inter-conal zone, which has a high temperature and a low flame background emission. Most of the visible, and much of the UV emission from the flame is associated with molecular species at the cooler, outer edges of the flame. In this secondary diffusion zone, combustion proceeds to completion using oxygen in entrained air. If a long path flame is used in FES

[21] G.F. Kirkbright, A. Semb, and T.S. West, *Talanta*, 1968, **15**, 441.

work, the detector 'sees' predominantly light from the inter-conal zone, and relatively much less of the unwanted light from the secondary diffusion zone. Thus the signal-to-background ratio is improved considerably.[20]

Kirkbright and colleagues[20,22] suggested an alternative approach to reducing the contribution from this unwanted molecular emission. They showed that, if the flame was enclosed inside a quartz tube a few centimetres in length, the secondary flame was lifted to the top of the tube (Figure 9B). Thus the inter-conal zone could be viewed with up to one hundred-fold reduction in flame background, especially in that from OH species. They further showed that a similar effect could be obtained by surrounding the flame with a stiff 'wall' of an inert gas such as nitrogen or argon.[20,23] The flame was then described as a 'separated' or 'inert gas-sheathed' flame.

## 12 Flame Photometry

While high flame temperature is an essential prerequisite to the sufficient thermal excitation of a wide range of elements to give useful emission intensities (see Chapter 1, equation 7), it should also be realized that some useful atomic emission determinations may be performed using low temperature flames. As discussed in Chapter 1, section 9, only elements with relatively very low excitation potentials, which therefore emit in the visible region of the spectrum, are excited in such flames. Because there are few such elements, those that do may be determined with a low incidence of spectral interferences. The flame used is usually air–propane or air–butane.

Figure 10 shows a schematic representation of a typical simple flame photometer. The number of wavelengths at which intense emission is observed is so small that a filter may be used in place of the monochromator invariably employed in FES. The use of an interference filter to isolate the wavelength of interest means that a large area of light emission may be viewed by the detector. As a consequence, a photomultiplier is unnecessary, and a much less expensive photoemissive or photodiode detector may be employed to convert the light signal to an electrical signal. Thus costs are heavily 'pruned' as a consequence of using a far cheaper detector, a much cheaper method of isolating the light of interest, and elimination of the need for a stabilized high voltage detector power supply. Thus flame photometers are simple, robust, compact, and inexpensive.

In environmental analysis, flame photometry is most widely used for the determination of potassium, which emits at 766.5 nm. It is also often used for the determination of sodium at 589.0 nm, although spectral interference problems (see Chapter 3) then may be encountered in the presence of excess calcium because of emission from calcium-containing polyatomic species. Molecular species are more likely to be found in cooler flames than in hotter flames. Some instruments use single, interchangeable filters, while others have three or more filters, for example for the determinations of potassium, sodium and lithium,

[22] G.F. Kirkbright, A. Semb, and T.S. West, *Talanta*, 1967, **14**, 1011.
[23] R.S. Hobbs, G.F. Kirkbright, M. Sargent, and T.S. West, *Talanta*, 1968, **15**, 997.

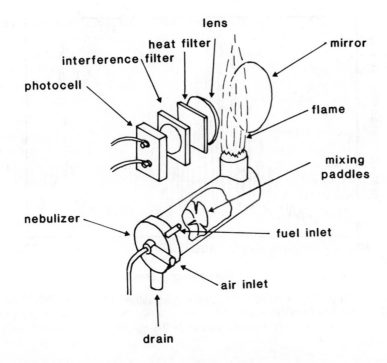

**Figure 10** *The essential components of a simple filter flame photometer*

incorporated in a filter wheel. The vast majority of instruments are single channel instruments, although some dedicated multichannel instruments had been built and were in routine use even 30 years ago.[24]

## Use of Molecular Emission from Cool Flames

Although molecular emission from flames is often regarded primarily as a nuisance in FES, it has also been exploited as an analytical technique in its own right for some determinations. For example, Figure 11 shows the emission spectra of $S_2$ and HPO species from low temperature, hydrogen–nitrogen–entrained air flames, which have been exploited for sulfur and phosphorus determinations. Many researchers have made use of $S_2$ emission-based flame photometric determination of sulfur and sulfur species,[25] although if the emission is obtained by spraying solutions into cool hydrogen flames, the interference problems associated primarily with different rates of formation of $S_2$ from diverse sulfur compounds are very severe.[26] Such problems are almost eliminated if species like sulfide or sulfite are converted to gaseous compounds such as hydrogen sulfide or

[24] A.M. Ure, *Br. Commun. Electron.*, 1958, **5**, 846.
[25] P.T. Gilbert, in 'Analytical Flame Spectroscopy', ed. R. Mavrodineanu, Macmillan, London, 1970, p. 181.
[26] R.M. Dagnall, K.C. Thompson, and T.S. West, *Analyst (London)*, 1967, **92**, 506.

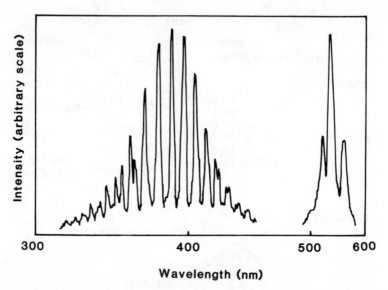

**Figure 11** *The emission spectra obtained when dilute sulfuric acid and phosphoric acid are nebulized into a hydrogen–entrained air flame. The spectrum on the left is that of $S_2$, and that on the right is that of HPO*

sulfur dioxide, and the latter are swept into the flame as pure gases: gaseous sample introduction procedures may be readily automated.[27,28]

The burner heads used in such cool flame emission studies are often simply quartz tubes. Figure 12 shows the burner system used by Arowolo and Cresser[27] for automated gas-phase sulfide determination, for example. Other species determined by cool flame emission techniques include chloride, bromide, and iodide, which give intense emission in the presence of indium.[29] The main application of cool flame emission techniques in environmental analysis is in speciation studies, for example for the separate determination of sulfite and sulfide, or as element-selective detectors in gas chromatography.

## 13 Molecular Emission Cavity Analysis

An alternative approach to producing intense molecular emission under cool flame conditions has been developed by Alan Townshend and colleagues at the Universities of Birmingham and, later, Hull.[30–33] These researchers found that intense molecular emission could be excited if samples were introduced into appropriate flames via a cavity in a moveable metal bar. The technique, which

[27] T.A. Arowolo and M.S. Cresser, *Microchem. J.*, 1992, **45**, 97.
[28] T.A. Arowolo and M.S. Cresser, *Talanta*, 1992, **39**, 1471.
[29] R.M. Dagnall, K.C. Thompson, and T.S. West, *Analyst (London)*, 1969, **94**, 643.
[30] M. Burguera, S.L. Bogdanski, and A. Townshend, *Crit. Rev. Anal. Chem.*, 1981, 10, 185.
[31] A.C. Calokerinos and A. Townshend, *Prog. Anal. At. Spectrosc.*, 1982, 5, 63.
[32] N. Grekas and A.C. Calokerinos, *Anal. Chim. Acta*, 1987, **202**, 241.
[33] N.P. Evmiridis and A. Townshend, *J. Anal. At. Spectrom.*, 1987, **2**, 339.

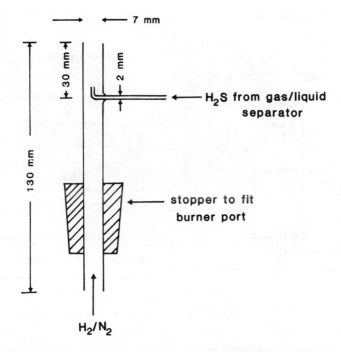

**Figure 12** *Simple burner head for use with an AutoAnalyser system for measurement of sulfide or sulfite via the emission of $H_2S$ or $SO_2$*

was applied primarily to a diverse range of non-metallic elements, became known as MECA. Although MECA spectrometers became commercially available, they do not appear to have found much routine application in environmental analysis.

## 14 Instrumental Requirements in AFS

Line sources are essential in AAS to obtain adequate sensitivity with a monochromator of only moderate resolution (see Chapter 1, section 7), to make sure that $I_t$ is substantially less than $I_0$. In theory, at least, modulated continuum sources may be used in AFS, because the readily isolated fluorescence emission following radiative excitation is characteristic of the element under investigation. For this reason, AFS is sometimes termed a 'self-monochromating technique'. However the radiative power from a continuum source which falls within the very narrow line width of a typical absorption profile is very small, even if the continuum source is very intense. Therefore detection AFS limits obtained with a source such as a 500 W xenon arc lamp are very poor.[34,35] Generally line sources, especially EDLs, are used as excitation sources (see section 3).

While in theory monochromators also are unnecessary in AFS, in practice most studies have been conducted using dispersive spectrometers such as the

[34] M.S. Cresser and T.S. West, *Spectrochim. Acta, Part B*, 1970, **25**, 61.
[35] M.S. Cresser and T.S. West, *Appl. Spectrosc. Rev.*, 1973, **7**, 79.

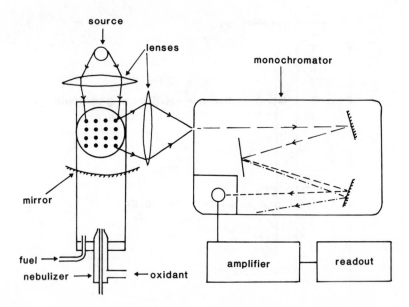

**Figure 13**   *Essential components of a typical dispersive atomic fluorescence spectrometer*

system shown in Figure 13.[36,37] The nebulizer and spray chamber are identical to those already described for flame AAS and FES, and typically the burner head will contain a circular or square array of holes, as in FES. The source will be modulated, as in AAS, to allow ready discrimination between the modulated fluorescence emission and unmodulated thermally excited emission from the flame. The monochromator serves to reduce the amount of flame background emission falling upon the detector, which would otherwise give excessively noisy (unstable) signals.

If the flame background emission intensity is reduced considerably by use of an inert gas-sheathed (separated) flame, then an interference filter may be used rather than a monochromator, to give a non-dispersive atomic fluorescence spectrometer as illustrated in Figure 14.[36-38] Noise levels are often further reduced by employing a solar blind photomultiplier as a detector of fluorescence emission at UV wavelengths. Such detectors do not respond to visible light. The excitation source is generally placed at 90° to the monochromator or detector. Surface-silvered or quartz mirrors and lenses are often used to increase the amount of fluorescence emission 'seen' by the detector.

Non-dispersive systems are very compact, and potentially less expensive than dispersive spectrometers. For this reason they may be used for multi-channel instruments, measuring upto six elements almost simultaneously.[39,40] Although

[36] I. Rubeska, V. Svoboda, and V. Sychra, 'Atomic Fluorescence Spectroscopy', Van Nostrand Reinhold, London, 1975.
[37] R.C. Elsed and J.D. Winefordner, *Appl. Spectrosc.*, 1971, **25**, 345.
[38] P.L. Larkins, *Spectrochim. Acta, Part B*, 1971, **26**, 477.
[39] R.M. Dagnall, G.F. Kirkbright, T.S. West, and R. Wood, *Anal. Chem.*, 1971, **43**, 1765.
[40] R.M. Dagnall, G.F. Kirkbright, T.S. West and R. Wood, *Analyst (London)*, 1972, **97**, 245.

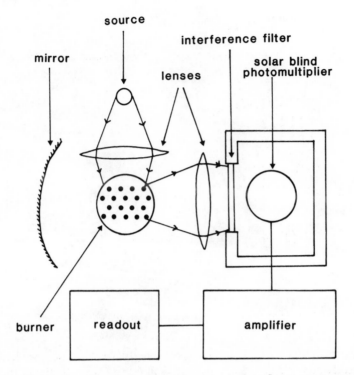

**Figure 14** *Essential components on a non-dispersive atomic fluorescence spectrometer*

such systems were put into commercial production in the 1970s, they were never widely used in environmental analysis.

In several of the early papers on AFS, non-premixed hydrogen flames were often employed, burning on total consumption burners (see, *e.g.* references 41 and 42). Although such flames had a low background emission, they did not give very stable signals, and they were not very efficient as atomizers, so that light from the source was often scattered by residual solid particulates, and even solution droplets, in the flame.[41] The resulting interferences were so severe that they soon fell out of favour, and must now be regarded as being of academic interest only.

Since source modulation and synchronous signal demodulation are practically essential in both AAS and AFS, and most other instrumental requirements are common to both techniques, it should come as no surprise to hear that the majority of AFS determinations have been performed using instrumentation designed primarily for AAS.[42]

[41] J.D. Winefordner and T.J. Vickers, *Anal. Chem.*, 1964, **36**, 161.
[42] M.S. Cresser and T.S. West, *Anal. Chim. Acta*, 1970, **50**, 5.

# CHAPTER 3
# Interferences and How to Overcome Them

From Chapters 1 and 2, it should be clear that flame AAS and AFS, and FES are all secondary analytical techniques which depend upon a comparison of signals from samples with those from standards. To yield accurate results, it is imperative in all three techniques that the determinant in samples and standards behaves in exactly the same way. If it does not, erroneous results will be obtained, and we say that an 'interference' has occurred. Interferences fall into four broad classes: physical, chemical, ionization, and spectral. Each of these classes needs to be considered in turn, together with the methods used to combat the problems which they would otherwise cause.

There are several stages in any flame spectrometric analytical procedure at which differences might occur between the behaviour of the determinant element in samples and in solutions. These include aspiration, aerosol generation, transport of aerosol droplets through the spray chamber, evaporation of solvent, volatilization of residual solid particulates in the flame, atomization, and the distribution and any subsequent recombination of atoms once formed. Interferences arising as a consequence of differences in behaviour during the first four stages are termed 'physical interferences'. Since the basic design of the instrumental components, the pneumatic nebulizer and spray chamber, which govern the sample or standard behaviour during these stages, is common to AAS, AFS and FES, it is logical to consider all three techniques at once, since both the incidence and extent of physical interferences should be similar in all three.

## 1 Physical Interferences

In flame spectrometry, physical interferences are related to transport of determinant from sample solution to the flame. The pneumatic nebulizer functions not only as a spray generator, but also as a pump.[1,2] Anything which influences the pumping rate will influence the size of the absorbance signal obtained. The pumping, or aspiration, rate is most sensitive to changes in viscosity of the sample solutions.

[1] I. Lopez Garcia, C. O'Grady, and M. Cresser, *J. Anal. At. Spectrom.*, 1987, **2**, 221.
[2] M.S. Cresser, in 'Sample Introduction in Atomic Spectroscopy', ed. J. Sneddon, Elsevier, Amsterdam, 1990, p. 13.

Aspiration rate is inversely proportional to viscosity. It is important therefore that sample and standard solutions have identical viscosities. This is best achieved by careful matrix matching.

It should be remembered in routine analysis of environmental samples that viscosity is highly temperature dependent, so that samples or standards should not be aspirated straight after removal from the fridge.[3,4] The viscosity effect is not as great as simple theory predicts, for two main reasons. As viscosity increases, and aspiration rate decreases, the energy of the nebulization gas is being distributed to less solution per unit time. This results in the production of a finer aerosol, which is transported with greater efficiency, partially offsetting the adverse effect of the increased viscosity.[3] The second reason is that the suction generated by a pneumatic nebulizer is strongly influenced by the aspiration rate.[4] If the rate of delivery of sample solution to the tip of the nebulizer capillary is too high, the nebulizer suction falls.[2] This too has a compensatory effect.

Aspiration rate is only a small part of the overall transport process in flame spectrometry. The production of aerosol and its transport through the spray chamber are also of great importance. The size distribution of aerosol produced depends upon the surface tension, density, and viscosity of the sample solution. An empirical equation relating aerosol size distribution to these parameters and to nebulizer gas and solution flow rates was first worked out by Nukiyama and Tanasawa,[5] who were interested in the size distributions in fuel sprays for rocket motors. Their equation has been extensively exploited in analytical flame spectrometry.[2,6-7] Careful matrix matching is therefore essential not only for matching aspiration rates of samples and standards, but also for matching the size distributions of their respective aerosols. Samples and standards with identical size distributions will be transported to the flame with identical efficiencies, a key requirement in analytical flame spectrometry.

Some organic solvents, for example isobutyl methyl ketone (4-methylpen-tan-2-one) and ethyl acetate, produce particularly fine aerosol and very high transport efficiency as a consequence.[8] Such solvents are therefore particularly useful for solvent extraction where very low detection limits are required.

Exact matrix matching is not always feasible; for example, the precise matrix composition may be unknown for various reasons. In such a case the standard additions method may be employed. The sample is spiked with at least two additions of known amounts of determinant in such a way that the matrix is not significantly altered, and the absorbance of spiked and unspiked samples is measured compared to that of aqueous standards, as shown in Figure 1. By extrapolation back to the negative extension of the concentration axis, the unknown concentration may be calculated.

It is clear that the standard additions method compensates for the change in slope of the calibration graph which the matrix components would cause. It is

[3] M.S. Cresser and R.F. Browner, *Anal. Chim. Acta*, 1980, **113**, 33.
[4] C.E. O'Grady, I.L. Marr, and M.S. Cresser, *Analyst (London)*, 1984, **109**, 1183.
[5] S. Nukiyama and Y. Tanasawa, *Trans. Soc. Mech. Eng., Jpn*, 1939, **5**, 68.
[6] B.L. Sharp, *J. Anal. At. Spectrom.*, 1988, **3**, 613.
[7] B.L. Sharp, *J. Anal. At. Spectrom.*, 1988, **3**, 939.
[8] M.S. Cresser, 'Solvent Extraction in Flame Spectroscopic Analysis', Butterworths, London, 1978.

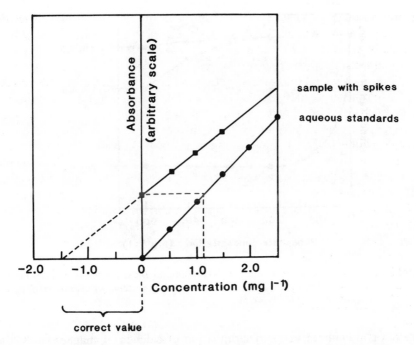

**Figure 1** *An illustration of the use of the standard additions method. From the aqueous standards calibration graph, the unspiked sample would appear to contain 1.13 mg of determinant per litre. The spiked samples allow a calibration graph with the correct slope to be used, and give a result of 1.50 mg $l^{-1}$*

also obvious that the standard additions method can only be used where the calibration graph is linear. For this reason it is best to make three or four standard additions in precise work. Small errors in slope from using too few additions may result in substantial errors in analytical results.

To use the standard additions method is time consuming compared to using the more direct comparison of samples and standards. If in doubt as to whether or not a physical matrix effect is likely to occur, and you have many samples to analyse, it may be worthwhile spending a little bit of time studying the effect of varying amounts of matrix components on determinant signal. Sometimes the range of variability of the matrix is such that approximate matrix matching may suffice, because the graph showing matrix effect on signal reaches a convenient plateau.

## 2 Chemical Interferences

Chemical interference arises when the determinant element forms a thermally stable compound with one or more concomitant ions or molecules present in the sample solutions.[9] The best known examples are the interference of phosphate,

[9] I. Rubeska and J. Musil, *Prog. Anal. Atom. Spectrosc.*, 1979, **2**, 309.

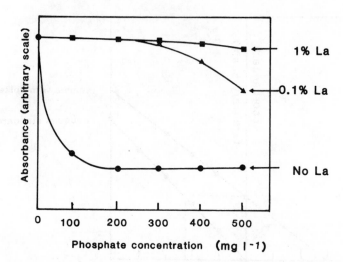

**Figure 2** *Effect of increasing amounts of phosphate on the absorbance from 20 mg l$^{-1}$ calcium in the absence of lanthanum, and in the presence of lanthanum at a final concentration of 0.1% or 1%*

silicate, or aluminium in the determination of calcium or magnesium in the air–acetylene flame. Figure 2, for example shows the effect of increasing amounts of phosphate on calcium determination.

There are two ways which are widely used to overcome this effect, to use a hotter flame or to add a releasing agent. A releasing agent is a substance which reacts preferentially with the interfering species, to form yet another thermally stable compound. Lanthanum or strontium are commonly used in the above example. The effect of lanthanum addition is also shown in Figure 2. Note that the amount of interference which may be tolerated depends upon the amount of releasing agent added. Note also that if the concentration of releasing agent is very high, a constant physical interference effect may be observed. This is not a problem, however, because the same amount of releasing agent is always added to all samples and standards, not only for matrix matching purposes, but also to compensate for the presence of trace amounts of determinant impurity in the releasing agent itself.

Because chemical interference depends upon the formation of thermally stable compounds, it is to be expected that the extent of the interference will be reduced if a hotter flame, such as nitrous oxide–acetylene, is employed. In many cases, unless the interferent and determinant concentrations are both high, the use of the hotter flame may be sufficient to prevent interference from occurring at all. Not all interferences are completely eliminated, however.

Chemical interferences are generally worse lower in the flame, and in cooler, fuel rich flames.[10] Indeed, making analytical measurements at two different heights is a useful procedure for investigation of whether or not chemical

[10] M.S. Cresser and D.A. MacLeod, *Analyst* (*London*), 1976, **101**, 86.

interferences are in practice occurring.[11] If they are not, the same answer should be obtained at both heights.

At a given interferent-to-determinant ratio, the extent of chemical interferences is invariably worse at higher determinant concentrations.[12] This is to be expected, because the small solid particles formed as the solvent evaporates will be larger and more difficult to vaporize and atomize at higher concentration.

## The Literature Interpolation Problem

One problem encountered by authors of textbooks on analytical techniques is knowing just how much information from original papers to include in their book. Sometimes, in the interests of brevity or presentational clarity, key information is inadvertently omitted. Thus, for example, an author would rarely say: 'With the particular nebulizer/spray chamber/burner head used by Bloggs, and at the fuel and oxidant flows Bloggs used, element X at 100 mg l$^{-1}$ did not interfere in the determination of element Y at 2 mg l$^{-1}$'. He is much more likely to say that Bloggs found that element X does not interfere in the determination of element Y. If he had used different conditions, or a different instrument, or even different concentrations, Bloggs might well have found serious interference problems. Thus take care when relying on textbooks.[12] To summarize, to minimize chemical interferences if they are suspected, work with dilute solutions and high in a fuel-lean flame, preferably nitrous oxide–acetylene.

# 3 Ionization Interferences

Atoms of some elements are relatively easily ionized at flame temperatures. This is particularly true for the alkali and alkaline earth elements, and other elements to the left of the periodic table. The first ionization potentials also tend to be lower for heavier elements within a particular group. For the group 2 elements, for example, ionization follows the order $Ba > Sr > Ca > Mg > Be$. This would not matter in flame spectrometric analysis, apart from a slight deterioration in sensitivity, if samples and standards were ionized to exactly the same extent. Suppose barium was to be determined in samples containing potassium, however. The potassium would be ionized:

$$K^0 \rightleftharpoons K^+ + e^-$$

This reaction would release a lot of electrons into the flame, which would suppress the ionization of barium, *i.e.*

$$e^- + Ba^+ \rightarrow Ba^0$$

Thus the presence of the potassium would enhance the signal of barium, to an extent which depended upon the potassium concentration, as shown in Figure 3. This is an example of ionization interference. It is an enhancement caused by

[11] A.A. Gilbert, I.L. Marr, and M.S. Cresser, *Microchem. J.*, 1991, **44**, 117.
[12] M.S. Cresser, *Lab. Pract.*, March, 1977, 171.

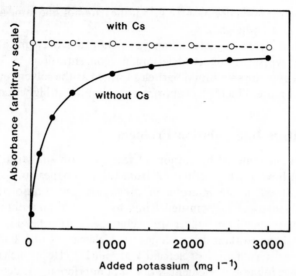

**Figure 3**  *Effect of increasing amounts of potassium on the absorbance, in a nitrous oxide–acetylene flame, of 2 mg l⁻¹ barium, in the presence and absence of caesium at a final concentration of 5 g l⁻¹*

depression of the determinant ionization by the presence of an easily ionizable concomitant element in a sample.

It is relatively easy to overcome ionization interferences. A large excess of an easily ionizable element such as potassium or caesium is added, which maintains the electron concentration constant. The substance added is known as an ionization buffer. Ionization interferences are, as might be expected, substantially worse in the nitrous oxide–acetylene flame than in the air–acetylene flame. It is a common misconception that an ionization buffer totally suppresses determinant ionization, but this is not strictly true. It buffers the degree of ionization at a fixed, reduced level.

# 4  Spectral Interferences

Whereas the incidence and extent of physical, chemical, and ionization interferences is similar in FES and flame AAS and AFS, spectral interferences are significantly different in the three techniques, and must be considered technique by technique.

## Spectral Interferences in Flame AAS

In Chapter 1, section 7, it was explained that very precise overlap of atomic absorption and emission profiles is required to obtain sensitive absorbance measurements. Absorption spectra of atoms at flame temperatures are much simpler than the emission spectra emitted by hollow cathode lamps. The possible transitions corresponding to electronic excitation of an atom may be shown as vertical lines on an energy level diagram, in which the vertical displacement

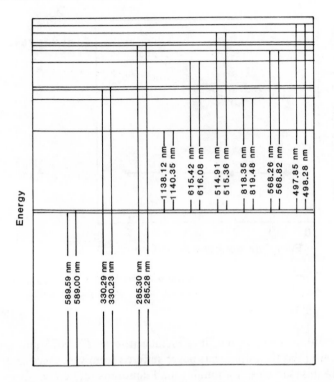

**Figure 4** *Partial energy diagram for sodium atoms, showing major absorption and emission wavelengths*

represents increasing energy difference between a pair of quantized energy levels [represented by horizontal lines (Figure 4)].

Not all transitions which are observed in the emission spectrum have the unexcited state (the ground state) as their lower energy level. In other words they require partial excitation before atomic absorption can occur. However, in flames, most atoms exist only in the ground state, and only transitions with the ground state as their lower energy state exhibit sensitive absorption.[13,14] Because the number of such transitions is small, the probability of overlap of the atomic absorption line profile of one element with the emission line profile of another element is extremely small. The spectral selectivity of AAS is therefore excellent in this respect.

Some examples of spectral line overlap are known.[15] For example, europium at 324.7530 nm interferes in the determination of copper at 324.7540, but europium does not interfere in copper determination at 327.3962 (see Figure 5). The fact that the interference occurs only at one analytical wavelength confirms that it is spectral in nature, since the extent of physical, chemical, or ionization interferences would be similar at all wavelengths.

[13] A. Walsh, *Spectrochim. Acta*, 1955, **7**, 108.
[14] M.S. Cresser, *Spectrosc. Lett.*, 1971, **4**, 275.
[15] J.D. Norris and T.S. West, *Anal. Chem.*, 1974, **46**, 1423.

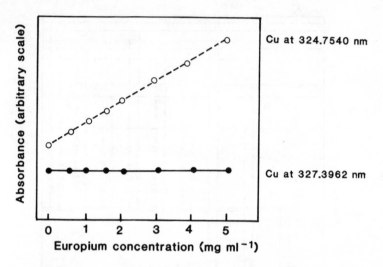

**Figure 5** *Effect of increasing amounts of europium on the apparent absorbance from 1 mg l⁻¹ copper at the two main copper resonance lines*

Other pairs of overlapping lines include iron at 271.9025 and platinum at 271.9038 nm (alas the author has never suffered from vast excesses of platinum in any of his samples, although small amounts do come from platinum crucibles!), silicon at 250.6899 and vanadium at 250.6905 nm, and aluminium at 308.2155 and vanadium at 308.2111 nm. Some further examples may be found listed in the useful monograph on spectrochemical analysis by Lajunen.[16] They do not present a problem as long as the analyst is aware of their existence. They may be circumvented by using a different spectral wavelength. Most of the above examples do not in fact occur at the wavelength corresponding to the best sensitivity.

At high concentrations especially, a number of elements produce significant concentrations of polyatomic species in flames. Such species absorb, and may therefore cause spectral interference. However the molecular absorption spectra are very wide compared with the atomic spectral lines. Figure 6, for example, shows how the presence of CaOH species in flames may interfere in the determination of barium by AAS. The formation of any solid particles in the flame causes scatter, which also causes an apparent broad band absorption, especially at lower wavelengths.

## Background Correction in AAS

Molecular absorbance and scatter are overcome by use of background correction techniques. In its simplest form, this may be achieved by measuring absorbance immediately adjacent to the determinant atomic absorption line. If light from a

[16] L.H.J. Lajunen, 'Spectrochemical Analysis by Atomic Absorption and Emission', The Royal Society of Chemistry, Cambridge, 1992.

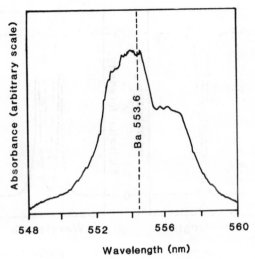

**Figure 6** *Part of the molecular absorption spectrum of calcium in an air–acetylene flame, showing the strong overlap with the main barium resonance wavelength*

continuum source is passed through the flame at the analytical wavelength, the molecular absorbance remains unchanged, but the atomic absorbance, as discussed earlier, is reduced to a negligible value. Thus subtracting the signal obtained with a continuum source from that obtained with a hollow cathode lamp provides a corrected absorbance.

In many modern atomic absorption spectrometers, this correction may be done automatically and simultaneously, time resolution of a few milliseconds being used to separate the two signals. Careful co-alignment of the two source beams is very important. To overcome this need, two other background correction systems have come into use over recent years, the Smith–Hieftje system and the Zeeman system.

In the Smith–Hieftje system, the lamp power is subjected to short pulses of high current.[17] This causes momentary bursts of high atom concentration in the hollow cathode. The emission line profile is broadened as an atom cloud forms just outside the cathode, which causes absorption at the centre of the emitted line profile, as shown in Figure 7. Effectively the single narrow emission line is split into a pair of lines immediately adjacent to the original line centre. Thus, in the normal mode, atomic and molecular absorption are measured, but in the pulsed mode, only molecular absorbance is monitored. The difference between the two signals provides a corrected atomic absorbance signal.

Zeeman background correction also depends upon line splitting, but in this instance most commonly the absorption line profile (the $\pi$ component) is split into two or more components (the $\sigma$ components) by the application of an intense

---

[17] S. Smith, R.G. Schleicher, and G.M. Hieftje, 'New Atomic Absorption Background Correction Technique', Paper 422, 33rd Pittsburgh Conference on Analytical Chemistry and Applied Spectroscopy, Atlantic City, 1982.

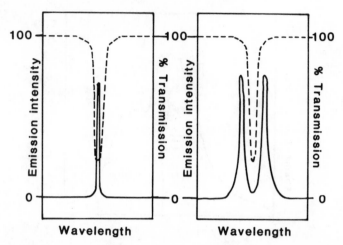

**Figure 7**  *Schematic representation of how the Smith–Hieftje background correction system works. On the left, the source emits a simple, sharp line at low current, and both atomic and molecular absorption would be measured. On the right, this simple line has effectively been split by a pulse of high lamp current into a pair of lines at either side of the atomic absorption profile, and only molecular absorption or scatter would be detected*

magnetic field.[18] Resolution of the $\pi$ and $\sigma$ components is achieved by exploiting the fact that the $\sigma$ components only absorb light with a particular plane of polarization. If the poles of an electromagnet are placed around the atomizer, and measurements are made alternately with the magnet off and on, atomic plus molecular absorbance and molecular absorbance only may be measured in turn. The difference between the two signals gives the corrected atomic absorbance. Figure 8 is a schematic representation of the processes involved. It is also possible to apply Zeeman splitting to the source atoms, which is more convenient in flame AAS, or to apply a constant field to the atom cell and to use a rotating polarizer to achieve separation of the $\sigma$ and $\pi$ components.

Zeeman and Smith–Hieftje background correction offer a number of advantages over continuum source background correction. They do not require precise co-alignment of two source beams and they allow a high degree of correction. They often work well even if the molecular absorption is highly structured. The Zeeman system described above uses conventional hollow cathode lamps at normal operating currents.

## Spectral Interferences in Flame AFS

In many respects the selectivities of AFS when an atomic line excitation source is used and AAS should be similar, in so far as both depend upon overlap of extremely narrow absorption and emission line profiles. However, there are differences in the extent of interference effects, even for resonance fluorescence,

[18]  R.D. Beaty, 'Concepts, Instrumentation and Techniques in Atomic Absorption Spectrophotometry', Perkin Elmer, Washington DC, 1988.

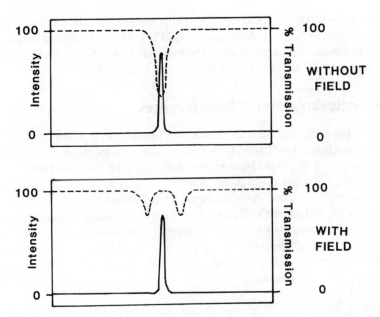

**Figure 8** *Schematic representation of how the Zeeman effect works. In the top sketch, no field is applied (normal AAS) and the absorption profile (---) overlaps the emission profile (—). In the lower sketch, with the field on, the absorption spectrum changes, and the σ components of the absorption profile no longer significantly overlap the source emission line profile. Broad band molecular absorption would still be detected, however*

because the fluorescence quantum efficiencies of the determinant and interferent elements may differ. For resonance fluorescence, the excitation and fluorescence emission are at an identical wavelength. However, part of the excitation energy may be lost by collisional deactivation prior to fluorescence emission (stepwise fluorescence) or after fluorescence emission (direct line fluorescence). Such emissions occur at longer wavelengths than the excitation wavelength. If stepwise or direct line fluorescence are measured, the prospect of spectral interference occurring is reduced even further in AFS compared to AAS. However, scatter problems are potentially more severe if the fluorescence technique is used, especially if the same resonance wavelength is used for excitation and emission. Scatter may be eliminated for many elements by measuring stepwise or direct line fluorescence, provided the emission from the source at the wavelength of the determination is filtered out prior to irradiation of the flame.

## Spectral Interferences in Flame AES

Spectral interferences are much more probable in AES than in either AAS or AFS, and it is important to be aware of the existence of nearby lines of concomitant elements which may be present in the sample solutions. Scanning the emission spectrum at high resolution in the immediate vicinity of the determination wavelength often will reveal potential spectral interferences. If in doubt, it is best to perform determinations at two or more different determinant analytical wavelengths, since it would be most unlikely that the extent of spectral interference would be identical at two or more wavelengths.

# 5  Conclusions about Interferences

Generally physical, chemical, and ionization interferences are similar in terms of incidence and extent in all three flame analytical atomic spectrometric techniques, but they are not a severe problem provided the analyst is aware of their existence, and takes the necessary precautions. Spectral interferences are not regarded as a serious problem in flame AAS or flame AFS, but are potentially much more serious in FES. Unless the analyst is certain that a particular FES determination is spectral interference-free for the samples in question, scanning and careful scrutiny of emission spectra from samples and standards is advisable, together with reliability checks using certified reference materials and/or determination at more than one wavelength.

# CHAPTER 4

# Optimization in Flame Spectrometry

In Chapter 3 we saw that many of the mechanisms which can result in interferences occurring are common to AAS, AFS, and FES. In this Chapter, we will see that for many of the parameters which need to be 'optimized' a similar approach again is employed in all three flame spectrometric techniques. However, there are also some noticeable differences between the techniques in this respect. Because there is a little more to do in AAS than in the other techniques, this technique has been discussed first, and then differences in AFS and AES are briefly considered.

## 1 What Do We Optimize?

The obvious answer to the above question is: 'The signal', but what does that answer mean? If some adjustment is made to a spectrometer which doubles the signal value obtained, but at the same time the noise on the baseline becomes twice as great also, the detection limit will remain unchanged. Thus if detection limits are to be optimized, then the maximum signal-to-noise ratio is required, and not the maximum signal. This simple and obvious fact is sometimes overlooked, partly because it is easier to maximize signals.

## 2 Optimization in Flame AAS

It should be realised by now that many components of the AA spectrometer are adjustable, and may need to be 'optimized'. It is helpful, not only to facilitate discussion here, but also as a memory aid in routine analysis, to divide the optimization parameters into three groups, namely:

(i) source-related parameters;
(ii) atomizer-related parameters; and
(iii) monochromator-related parameters.

We shall briefly consider each of these three groups in turn.

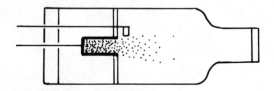

**Figure 1**   *Schematic representation of the cloud of cool atoms which escapes from the centre of the hollow cathode at high lamp operating currents*

## Source-related Parameters

### Effect of Lamp Current

Source operating power may have a substantial influence upon both absorbance and upon signal-to-noise ratio. For hollow cathode lamps, the maximum safe routine operating current is often printed on the lamp label. If this current is exceeded for more than a very short period, the lamp may be permanently damaged. Normally a much lower current of around 4–7 mA is used in routine work. Consider what happens to atoms in the immediate vicinity of the hollow cathode if the lamp current is excessive (Figure 1). Sputtered atoms will escape from the immediate vicinity of the hollow cathode.

Now try to imagine what effect these atoms might have upon the shapes of resonance lines emitted by the hollow cathode lamp. The atoms freed from the cathode discharge are cooler and will be in the ground state. They will therefore give very sharp absorption bands. Absorption will be greatest at the centre of the resonance lines emitted by the lamp. The line profile from the lamp thus becomes broader, and eventually will exhibit self absorption, and ultimately self reversal, giving the appearance of two lines close together immediately to either side of the original line. The latter effect is similar to that already discussed in the section on Smith–Hieftje background correction in Chapter 3.

The absorption line profile in the atomizer (*e.g.* in the flame) will still peak at the initial emission line peak. Absorbance will be reduced as the emission line becomes broader, and even more dramatically when the emission line shows reversal. Thus atomic absorption signal decreases with increasing lamp current (Figure 2). As might be expected, the drop off in signal is greater for more volatile elements such as cadmium and zinc.

The biggest signal is obviously obtained at very low current. But will very low current necessarily give the best precision? At very low currents, the discharge may become less stable, and also a higher photomultiplier or amplifier gain will be needed, so signal-to-noise ratio may become poorer at very low current (Figure 3). In Figure 3, the noise problem is readily visible, because the absorbance signal has been plotted continuously as a function of time using a chart recorder. The effect is less readily apparent if a digital readout is the only readout available. It is then necessary to calculate and compare relative standard deviation (RSD) values under diverse conditions.

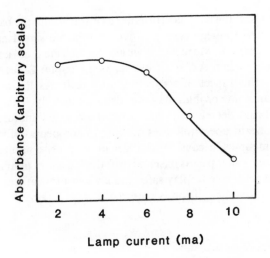

**Figure 2**  *Effect of increasing hollow cathode lamp current upon the absorbance from 1 mg l$^{-1}$ of cadmium*

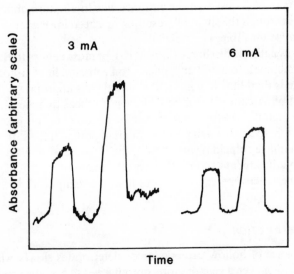

**Figure 3**  *Effect of lamp current on the normalized signal from 0.5 and 1.0 mg l$^{-1}$ of zinc in an air–acetylene flame*

## Effect of Lamp Warm Up Time

Considering the complex nature of the hollow cathode lamp, it should not be expected that the hollow cathode lamp output will become completely stable within a few seconds of the lamp being switched on. In practice, lamps take a few minutes (about five is usually sufficient) to stabilize. It is instructive to think about what effect this warm up period will have upon baseline and atomic absorption signals. If the lamp gets brighter as it warms up, the baseline will tend to drift below zero (remember, a positive absorbance signal means less light reaching the detector). More light would be detected as less absorbance. However a double beam AA spectrometer (see Chapter 2, section 10) should compensate for this aspect of source intensity drift. As the lamp warms up, emission resonance line profiles may broaden significantly. This may result in a small but significant decrease in signal over the first few minutes.

Even double beam operation does nothing to compensate for this aspect of signal drift. Sometimes, because of changes in the precise nature of the hollow cathode discharge, other trends in signal drift during warm up may be observed. It is best as a general rule to play safe, and always allow at least 5 minutes for lamp warm up if good precision is required.

## Lamp Alignment

If the lamp is properly aligned, light from the most stable, central region of the cathode is focused on the monochromator entrance slit. If the hollow cathode lamp is not carefully lined up with the slit, light from the edge of the discharge (less stable) may be focused on the entrance slit. In addition, much useful light may not pass through the slit at all, resulting in a need for higher electronic gain. Both may cause unstable, noisy signals.

Most hollow cathode discharges appear red, because neon is the most common filler gas. Sometimes, to avoid unwanted neon emission lines, argon is used, and the discharge is then lilac in colour. It is not always appropriate to use the red image of the hollow cathode produced by a lens to check that the lamp is properly aligned, although it is useful for approximate alignment. This is so because UV and red light will focus through a lens at different points. If the cathode is off axis, when the red image coincides with the slit, the UV 'image' may be to one side of the slit. Fine adjustments should therefore be made using the meter or digital readout of the spectrometer to optimize light throughput.

## Lamp Deterioration

The performance of hollow cathode lamps deteriorates slowly with use. After several months, or even a year or more, output tends to become progressively less stable and/or less intense. Because the loss in precision is gradual, it may well pass unnoticed. It is therefore useful to keep a recorder trace of signal stability after a new lamp has been in use for a few hours for comparison purposes. To be useful, however, it is important to have made a note of the slit width, photomultiplier gain setting, wavelength, and lamp current used, and the lamp position must of course have been carefully optimized. Some analysts prefer, on grounds of simplicity, to make a note of the absorbance attainable under optimized flame conditions from a specified determinant standard. However it must then be remembered that a decline in this parameter may be related to nebulizer deterioration rather than lamp deterioration.

## Choice of Lamp

Fairly new lamps from the same manufacturer, and even some lamps from different manufacturers, tend to give rather similar performance. This reflects the strict quality control procedures observed during lamp manufacture. Multi-element hollow cathode lamps are available for some combinations of elements, but, in the author's experience, they tend to give rather inferior performance to that of single element lamps. If sources such as microwave or electrodeless discharge lamps are employed, these too will have different performance characteristics. When comparing sources, the absorbance signal size and stability for a suitable dilute standard solution should be compared, with all other conditions carefully optimized, and not just the lamp intensities and stabilities. The most intense lamp may not be the one which gives the greatest signal or signal-to-noise ratio.

## Atomizer-related Parameters

### Choice of Atomizer

We have already seen in Chapter 2 that choice of atomizer system to be used may have a dramatic effect upon sensitivity, and thus upon signal-to-noise ratio. It is necessary to choose not only between flames, electrothermal atomization (ETA), and cold vapour and hydride generation techniques (which are discussed in Chapter 6), but sometimes also between different flames. Those elements which tend to form thermally stable oxides, such as Al, Ti, Si, Zr, may only be determined in a hotter, reducing nitrous oxide–acetylene flame. They cannot be determined with useful sensitivity in the air–acetylene flame. Some elements, Ba and Cr for example, may be determined in air–acetylene, but are more efficiently atomized in nitrous oxide–acetylene.

### Effect of Fuel-to-oxidant Ratio

For some elements, especially those which tend to form thermally stable oxides, fuel-to-oxidant ratio may have a dramatic effect upon atomic absorbance signal. Figure 4, for example, illustrates the effect of increasing fuel flow upon aluminium determination.

Precision is much poorer for such determinations if the fuel flow is not optimized, because a higher degree of scale expansion is required. This shows clearly in the recorder traces in Figure 5.

One point which is often overlooked when optimizing fuel-to-oxidant ratio is that the optimum fuel flow is sometimes matrix-dependent. For example, the determination of calcium in water using an air–acetylene flame is more sensitive if a fuel-rich flame is used. If, however, the samples contain dilute sulfuric acid, a more fuel-lean flame usually gives substantially improved sensitivity. As a general rule, it is best to find optimal conditions for the particular matrix which you are analysing.

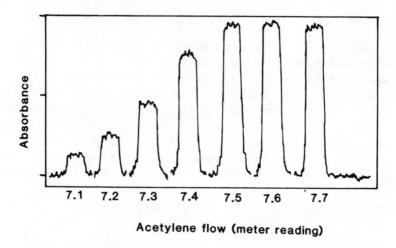

**Figure 4**  *Effect of increasing acetylene flow on the absorbance from 10 mg l⁻¹ of aluminium in a nitrous oxide–acetylene flame*

## Optimization of Burner Position

Signal and signal-to-noise ratio depend upon the passage of the light beam from the lamp through the flame centre at the optimum height. Thus the burner position must be optimized in three respects:

   (i)  height;
   (ii)  sideways (lateral) position; and
   (iii)  rotation.

Signal and signal-to-noise ratio often vary markedly with height of observation, and again the optimal height is often dependent upon the matrix components present in the sample. Optimal height should therefore be established for each matrix type. Lateral position is important because passage of the light beam through the edge of the flame, rather than the flame centre, generally results in smaller signals and poorer signal stability. The latter is partly due to the fact that much of the molecular thermal emission from the flame (the flame background colour) occurs from the cooler, outer edges of the flame, resulting in noisier photomultiplier signals if this flame zone is viewed. If the burner head is not rotationally aligned, the light beam may not pass through much of the atom population in the flame at all.

Sometimes burner rotation is recommended as a method of losing sensitivity deliberately in flame AAS, to extend the linear range of the calibration graph upwards and to thus avoid the need for sample solution dilution when many samples are to be analysed. Such a practice is not without its drawbacks. The flame edges make a greater relative contribution than the flame centre to the total signal, so that precision is generally reduced. Moreover matrix interferences tend

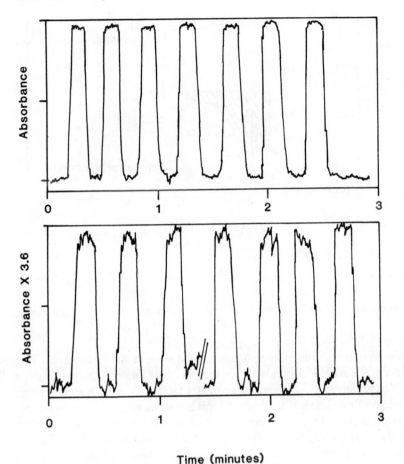

**Time (minutes)**

**Figure 5** *Effect of operating at optimal (top) and sub-optimal (bottom) fuel flow when repeatedly nebulizing 10 mg l⁻¹ aluminium into a nitrous oxide-acetylene flame*

to be worse at higher determinant concentrations, as mentioned in Chapter 3. Dilution is better avoided by employing a small impact cup in place of the impact bead, if possible.[1,2] This device (Figure 6) reduces transport efficiency to *ca.* 0.5% by dumping the larger aerosol droplets, and thus reduces the extent of interferences.[1]

## Burner Design, Warm Up, and Cleanliness

It is appropriate to consider these three aspects together, because they are inter-related to a certain extent. Often the analyst has no say in burner design;

[1] M.S. Cresser, *Analyst (London)*, 1979, **104**, 792.
[2] C.E. O'Grady, I.L. Marr, and M.S. Cresser, *J. Anal. At. Spectrom.*, 1986, **1**, 51.

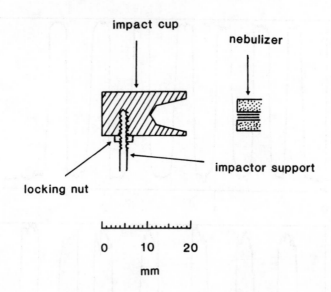

**Figure 6**  *The design of an impact cup for use in analytical flame spectrometry*

he/she has to use the burner heads provided with the instrument. However burner heads with a triangular cross section in the immediate vicinity of the slot are to be preferred, because such a geometry facilitates smooth air entrainment to the edges of the flame. If the burner head is flat, there is severe turbulence at the point of air entrainment where the inflowing air has to change direction very sharply. This turbulence results in localized heating of the burner head, which adversely effects both warm up time and the tendency of the slot to clog (see Chapter 2, section 5). If a flat-topped head must be used, try to avoid nebulization of concentrated salt solutions.

If the burner head does clog, the flame starts to appear uneven, and signal stability deteriorates rapidly. It is necessary to extinguish the flame, and clean the slot. Allow the burner head to cool first. Use of a razor blade with air flowing through the slot is sometimes adequate, taking care not to burn the fingers, although thorough flushing with distilled water and drying may be necessary for a longer lasting cure.

It takes 5 to 10 minutes for a burner head to warm up thoroughly, depending upon the burner design. Figure 7 shows what happens for 10 mg l$^{-1}$ aluminium, by both absorption and emission, if the lamp is allowed to warm up but not the nitrous oxide–acetylene burner head. It should be noted that double-beam operation will do nothing to prevent this period of instability. On many instruments changes in the appearance and flame geometry of the nitrous oxide–acetylene flame over this period are clearly visible. Figure 8 shows how much better the results are if both the burner head and, for AAS, the hollow cathode lamp, are allowed to warm up for 10 minutes prior to the nebulization of samples or standards.

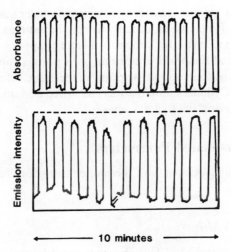

**Figure 7** *The effect of inadequate burner head warm up time on the absorbance (top) and emission (bottom) signals from repeated nebulization of a 10 mg l⁻¹ aluminium solution*

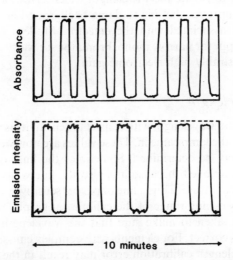

**Figure 8** *Beneficial effect of allowing a 10 minute warm up period, compared with the conditions in Figure 7*

## Gas Flow Stability

Fuel and air or nitrous oxide flow stabilities must be adequate for good precision. Intermediate balast tanks help to smooth out fluctuations caused by compressors.[3] As mentioned in Chapter 2, section 5, nitrous oxide cools when it is subjected to a sharp pressure drop, which results in cooling of the cylinder head, and sometimes in instability. The effect is not as important if the nitrous oxide operating pressure

[3] M.S. Cresser and G. Wilson, *Lab. Pract.*, Feb. 1973, 117.

is at around 45 p.s.i. or above, but may be severe on older instruments operating at lower pressures. It may be reduced by using a balast tank at an intermediate pressure, say 60 p.s.i.,[3] or by strapping heating tape to the cylinder head. Acetylene is supplied dissolved in acetone on a porous support. The cylinder should be allowed to attain normal working temperature before use, especially in very cold weather. Cylinders should always be stored upright, to prevent acetone solutions from entering the fuel gas lines.

## Monochromator-related Parameters

The main variables associated with the monochromator are the slit width and wavelength setting.

### *Choice of Slit Width*

The slit width, or spectral band pass, chosen depends largely upon the complexity of the source spectrum being used. If the determinant element has a complex spectrum in the immediate vicinity of the resonance line to be used, a narrow slit must be employed to isolate the wavelength which gives the best sensitivity. A narrow slit also improves the line-to-flame background emission ratio, which may reduce noise associated with the detector. Modern lamps usually have sufficient light output for narrow monochromator slit widths to be used without the need for undesirably high electronic gain.

### *Choice of Wavelength*

Usually the optimum wavelength with respect to sensitivity is written on the lamp label or on the box the lamp came in. These values are now well known for all the elements which may be determined. Sometimes the line which gives the best sensitivity is not employed, because it is also the line which gives the greatest calibration graph curvature. Thus a line of poorer sensitivity may be preferred to extend the linear working range of the calibration graph.

Care should be taken to make sure that the wavelength calibration of the monochromator is correct. For elements with complex emission spectra, such as iron, a small wavelength calibration error may result in the wrong wavelength being employed accidentally. This may seriously adversely effect sensitivity and precision.

If in doubt, check that the wavelength thought to be the one giving optimal sensitivity does indeed do so. Alternatively check the wavelength calibration in that vicinity using a hollow cathode lamp of an element with a much simpler, and thus unambiguous, spectrum in the region of interest.

## 3   Optimization in Flame AFS

There are many similarities in the optimization procedures used in AAS and AFS, but also there are some significant differences. The latter arise as a

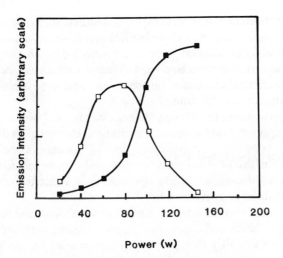

**Figure 9**    *Effect of increasing applied microwave power upon the source intensity (■) and zinc atomic fluorescence intensity (□), using a vacuum-jacketed zinc EDL source*

consequence of the different instrumental configuration used, as discussed in Chapter 2, and from the self-monochromating nature of AFS compared to AAS. However, it is still convenient to consider optimization under the same three sub-headings, source-related parameters, atomizer-related parameters, and monochromator-related parameters. To avoid unnecessary repetition, only factors which differ in AFS compared to AAS are considered below.

## Source-related Parameters

### Lamp Operating Parameters

We saw earlier in this chapter that, in AAS, increasing the current through a hollow cathode lamp results in line broadening of the emitted resonance lines, and an associated deterioration in sensitivity. A similar decline in sensitivity with increasing power is also observed when microwave- or r.f.-powered EDLs are used. In AFS the situation is quite different. Provided the width of the broadened emission line profile is still smaller than that of the absorption line profile, more energy will still be absorbed as the power supplied to the source is increased, and the fluorescence signal will increase with increasing applied power. Eventually, however, the source emission line profile will become broader than the absorption line profile, and will also exhibit line reversal. When this situation occurs, much of the additional energy emitted by the source can no longer be absorbed by the determinant atoms. In the most severe cases, little or no overlap will occur between the emission line profile and the absorption profile, and little or no fluorescence will therefore be observed. This situation is represented schematically in Figure 9.

From the above discussion it should be clear that optimization of source operating conditions can only be effectively achieved by measuring the effect of source current or power upon the fluorescence signal from a standard determinant solution, and not from the effect upon direct scattered or reflected light from the source. Consideration must also be given to effects upon fluorescence signal-to-noise ratio, rather than simply to effects on signal size, and to possible adverse effects upon source operational lifetime.

While conventional or high-intensity (boosted output)[4] hollow cathode lamps are usually simply operated at room temperature, electrodeless discharge lamps are sometimes cooled with a regulated flow of air maintained at a constant temperature,[5] and this flow too must be optimized with respect to signal-to-noise ratio. Sometimes these sources are operated in a vacuum jacket to enhance sensitivity and/or to improve stability.[6]

Perhaps the most marked difference between flame AAS and flame AFS is the fact that, in the latter technique, the signal increases with the useful source intensity. Source intensity (for a constant line profile) has no effect in AAS, because absorbance is a ratio (see Chapter 1). For this reason, over almost three decades a great deal of research effort has gone into trying to produce more intense and stable sources for use in AFS. For some elements, very low detection limits have been obtained using lasers as excitation sources. A comparison of detection limits by AFS using diverse sources may be found in the useful critical and comprehensive review of AFS by Omenetto and Winefordner.[7] Such sources have found very limited application in routine environmental analysis, primarily because of cost and lack of standard commercially available instrumentation; they will not be considered further here.

## Lamp Alignment

Lamp alignment may be a little more difficult in AFS than in AAS, depending upon the type of source and optical configuration employed.[7,8] As for source operating power optimization, it is more reliable to use a genuine fluorescence signal rather than reflected light for optimization of source alignment.

## Atomizer-related Parameters

Both AAS and AFS with flame atomizers depend upon the stable and reproducible production of ground state atoms. Therefore atomizer parameters such as fuel-to-oxidant ratio or burner head position relative to the excitation beam will be equally important in both techniques. The major difference in AFS lies with the types of flames sometimes employed.

In flame AAS, the light emitted from the source at the resonance wavelength is usually substantially more intense than the light emitted from the flame at the

[4] J.V. Sullivan and A. Walsh, *Spectrochim. Acta*, 1965, **21**, 721.

[5] R.F. Browner, *Analyst (London)*, 1974, **99**, 617.

[6] K.E. Zacha, M.P. Bratzel, J.D. Winefordner, and J.M. Mansfield, *Anal. Chem.*, 1968, **40**, 1733.

[7] N. Omenetto and J.D. Winefordner, *Prog. Anal. At. Spectrosc.*, 1979, **2**, 1.

[8] I. Rubeska, V. Svoboda, and V. Sychra, 'Atomic Fluorescence Spectroscopy', Van Nostrand Reinhold, London, 1975.

same wavelength and falling within the spectral bandpass of the monochromator. Thus the flame emission generally contributes little to the signal noise in flame AAS. In AFS, on the other hand, the fluorescence signal, especially at concentrations near the detection limit, may be appreciably smaller than the flame emission signal. Even although synchronous demodulation (see Chapter 2, section 4) is used to improve discrimination between the two signals, the flame background contributes considerably to the signal noise level. Thus the use of low-background flames such as separated flames (Chapter 2, sections 11 and 14) is advantageous in AFS, in which case both the fuel-to-oxidant ratio and the separator gas flow must be carefully optimized.

As an alternative to separated flames for reducing flame background, hydrogen may be used as a fuel rather than acetylene, although care is then needed to make sure that chemical interferences do not start to occur. In the early days of AFS, often improving detection limits became an end in itself, with little or no thought being paid to potential interference problems.

The intensity of fluorescence emission depends upon, among other factors, the stability of the radiatively excited determinant atoms. The presence of a large number of polyatomic species in flame atomizers, however, favours collisional deactivation over fluorescence emission. Thus a number of workers have employed argon/oxygen/hydrogen flames in AFS in preference to air–acetylene flames.[8,9] Optimization of the three gas flows must then be performed, bearing in mind the need to make sure that interferences do not become a problem and that flow velocity must adequately exceed burning velocity. Again it should be stressed that these flames have rarely been used in environmental laboratories for routine analysis.

## Monochromator-related Parameters

In AFS there is no need to isolate a single wavelength in the fluorescence emission spectrum from nearby, less-intense emission wavelengths, since all lines contribute to the fluorescence signal. Therefore quite large spectral bandpasses are often employed in flame AFS, especially when a low-background flame is being used. Indeed, as seen in Chapter 2, section 14, non-dispersive, filter-based systems may sometimes be employed.[7,8]

# 4 Optimization in Flame AES

## Atomizer-related Parameters

In some respects, optimization procedures in flame AES are simpler than those in either AAS or AFS. This is so because of the absence of a light source (apart from the flame atomizer). Moreover, there is little incentive to employ hydrogen as a fuel in flame AES, because the lower flame temperatures (compared to the corresponding acetylene flames) do not favour intense thermally excited atomic emission. Air–acetylene and nitrous oxide–acetylene flames are most widely

[9] D.R. Jenkins, *Spectrochim. Acta, Part B*, 1970, **25**, 47.

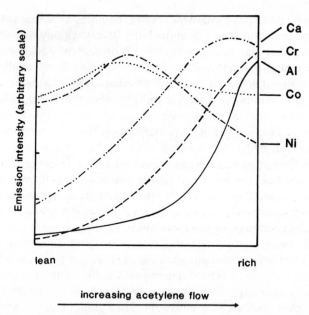

**Figure 10**   *Effect of increasing acetylene flow on the emission intensities from a nitrous oxide–acetylene flame for a selection of elements*

employed. Fuel-to-oxidant ratio must be very carefully optimized, because of the need to optimize both free atom production and flame temperature. Figure 10 illustrates some typical curves showing fuel flow effects for a selection of elements in nitrous oxide–acetylene. Burner height must also be very carefully optimized, as must be the rotational and lateral positions of the burner head if a long-path burner head is used.

It should be remembered than optimization should be with respect to signal-to-noise ratio rather than to the size of the signal.

As in all flame spectrometric techniques, care should be taken to make sure that the nebulizer performance, if the nebulizer is adjustable, is optimized. This may involve direct adjustment of the nebulizer capillary position or minor adjustments to the position of the impactor, if fitted.

In flame photometry, there is little scope for atomizer optimization, because of the simplicity of the instrument design. However, fuel flow should be carefully adjusted.

## Monochromator-related Parameters

The function of the monochromator in AES is to isolate the determinant spectral wavelength of interest from the emission from all concomitant matrix emitting elemental or molecular species. This frequently means that a narrow spectral bandpass must be selected. It is however generally slightly easier to make sure in AES than in AAS that the optimal wavelength is being employed since emission spectra often may be scanned directly.

If the flame background around the wavelength to be used is variable, as a consequence of variable amounts of potentially interfering matrix elements, it is advisable to scan the emission spectrum in the vicinity of each wavelength of interest for each sample and standard in routine analysis. Spectral scan rate, the wavelength interval studied for each determination, and electronic damping must all then be carefully optimized.

# 5 How Careful Do I Need To Be?

The magnitude of performance deterioration for failing to optimize any one of the steps described in sections 2 to 4 varies enormously, and from element to element. How painstaking you should be when setting up for a particular determination depends upon how close to the detection limit you will be working, and upon the degree of precision required. With experience, many shortcuts may be deemed perfectly acceptable. Where particular care is needed with optimization, a warning is provided in the element-by-element section of Chapter 7.

# CHAPTER 5

# *Sample Preparation*

The vast majority of environmental analyses completed by flame spectrometry either involve direct analysis of aqueous samples or analysis of solid samples after sample dissolution. It is appropriate at this point, therefore, to consider briefly the implications which the requirement to have the sample in solution form has in flame spectroscopic analysis, and that is the prime purpose of this chapter. However, it is also important never to lose sight of the fact that appropriate sampling and sub-sampling techniques are a crucial prerequisite to the generation of meaningful environmental data. The analytical process often starts in the field, and that is the stage at which we should begin to look at sample preparation.[1]

## 1  Field Sampling Technique

A sample taken for analysis must be representative with respect to a specific question being asked. If the material being sampled has uniform chemical, physical, and biological properties, taking a sample may be very straightforward; it simply involves removing part of the whole without introducing any contamination, regardless of the question. Even in this simplest case, the latter constraint may impose problems for hard materials such as some minerals. Environmental materials are rarely homogenous, often varying in three dimensional space and time.

   Consider, by way of an example, the sampling of an upland river. Water may be draining into the river from several different depths in the surrounding soil or bedrock, or from several different soil types. Water from each source may have its own solute chemical characteristics. The relative importance of diverse hydrological pathways may change seasonally or with climate even over a few hours in a heavy rain storm event.[2,3] The river water solute composition will thus change over the same timescales. Water entering rivers is often substantially enriched in carbon dioxide from the soil atmosphere compared to water equilibrating with the above ground atmosphere. Out-gassing of this excess carbon dioxide may result in a pH rise of upto two pH units.[2] Thus even an

[1] M.S. Cresser, *Anal. Proc.*, 1990, **27**, 110.
[2] M.S. Cresser and A.C. Edwards, 'Acidification of Freshwaters', Cambridge University Press, Cambridge, 1987.
[3] A.C. Edwards, J. Creasey, and M.S. Cresser, in 'Hydrochemical Balances of Freshwater Ecosystems', IAHS Publication No. 150, Oxford, 187, 1984.

apparently simple sample material like river water is highly dynamic. A hundred or more samples may be needed to reasonably characterize even a small river if spatial and temporal variability are of interest.

Other natural water samples are also highly variable. Rainwater chemical composition varies markedly from storm to storm at many sites, especially in smaller countries (land masses) with maritime climates, and often also shows marked changes even within individual precipitation events.[4] Lake water may exhibit seasonal trends and even diurnal trends. The latter tend, as a consequence of effects of photosynthesis, to be especially pronounced in shallow, eutrophic lakes in regions with warm climates.[5] Larger, deeper lakes in cooler regions tend to be better buffered against rapid change, as are the oceans, but they may exhibit chemical gradients as a consequence of thermal stratification.

Sampling suspended materials in rivers and lakes is particularly problematic. Not only does the suspended load tend to vary in space and time, but also it is necessary to take care not to introduce systematic error when taking the sample. For example, sampling devices which suck or pump water from depth along pipes can result in substantial loss of suspended materials by sedimentation. It is better therefore to have a sample bottle which may be both opened and then closed at the appropriate depth.

Plant samples too are highly variable in space and time, and the concentrations of some key plant nutrient elements may vary by three-fold or more over a period of a month or two. Moreover, for evergreen perennial species, there may be a problem in sampling current leaf growth rather than older foliage. Plant sampling strategies must therefore be carefully thought out and rigidly adhered to.

The total element contents of soils and rocks tend to change significantly only over relatively long time scales (decades to centuries), unless they are subjected to substantial pollution loads. However, the much smaller pools of biologically available nutrient elements in soils tend to be more dynamic, especially during periods of active growth. Speciation in polluted soils may change substantially over days or even hours following a serious pollution episode.[6]

This section is not intended to give a comprehensive account of environmental sampling problems, but rather to emphasize the importance of giving the necessary thought to sampling technique. Because many environmental samples are biologically active even when handed into the analytical laboratory, they must be stored and handled accordingly, especially if speciation studies are to be performed rather than total element analysis.

## 2   Sample Treatment and Storage

### Waters

Water samples are best stored cold, at 1–4 °C, but they should not be frozen. Freezing sometimes results in irreversible precipitation reactions occurring.

[4] A.C. Edwards and M.S. Cresser, *Water Air Soil Pollut.*, 1985, **26**, 275.
[5] M.S. Cresser, A.C. Edwards, and Z. Parveen, *Endeavour*, 1993, **17**, 127.
[6] Z. Parveen, A.C. Edwards, and M.S. Cresser, *Sci. Total Environ.*, in press.

Samples are often acidified with a small amount of nitric or hydrochloric acid if to be used for trace metal determinations, and stored in plastic rather than glass vials to minimize the risk of adsorption losses. Storage containers should be acid washed and thoroughly rinsed with deionized water. Acidification may result in precipitation of naturally occurring dissolved organic matter, however. Speciation studies should be completed immediately after collection on fresh samples.

## Plant and Other Biological Tissues

One of the first decisions which must be taken is whether or not to wash the samples. The answer to this question depends upon what is to be determined. Some elements, especially magnesium, potassium, and manganese, may be leached slightly from leaves even by a distilled water rinse. Larger amounts are leached by an acid rinse, even if the acid is very dilute. This suggests that washing is undesirable. On the other hand, short growing plants, and especially those with hairy foliage, may collect significant amounts of soil particles as a result of water splash. For elements such as aluminium and iron, which are present in much larger amounts in soil than in plant tissues, washing may be necessary to reduce this contamination. Excess water is removed by shaking, and often by dabbing with low contamination risk tissues; the risk can be ascertained by a simple leaching experiment.

Root samples must be washed to reduce contamination, but even so complete clean up is impossible for many soil types. Sometimes a brief period of ultrasonic treatment is used in a water bath. Reliable studies of leaching losses from plant roots are scarce because they are so difficult to conduct, and the results are rarely unequivocal.

Unless unstable organic components such as plant pigments are to be determined, plant tissues are invariably oven dried prior to analysis. Large samples are usually dried on shallow, open trays in forced draught ovens. Care is needed to avoid contamination from brass or galvanized iron trays. The samples are then ground, again avoiding contamination, and smaller sub-samples further dried over night at 80 °C or for an hour at 105 °C prior to dry ashing or wet digestion. Dried samples are usually stored in sealed glass vials or small polythene bags.

## Soils

Very few determinations are usually conducted on field-moist soils, although the practice is essential for the determination of extractable ammonium-N and nitrate-N.[7,8] These two species may change very rapidly in biologically active soils, especially if the soils are dried and rewetted. However, these determinants are not measured by flame spectrometry. Speciation studies are best performed

---

[7] I.L. Marr and M.S. Cresser, 'Environmental Chemical Analysis', Blackie and Son, Glasgow, 1983.

[8] D.R. Keeney, in 'Methods of Soil Analysis: Part 2—Chemical and Microbiological Properties', 2nd Edn., ed. A.L. Page, R.H. Miller, and D.R. Keeney, American Society of Agronomy, Madison, WI, 1982, p. 711.

on hand-sorted and 2 mm-seived fresh moist soils, because even air drying may result in substantial changes in the species distribution of manganese, iron, and, to a lesser extent, potassium. If field-moist soils are analysed, separate sub-samples must be oven dried to allow results to be expressed on an oven-dry basis.

Soils are usually air dried in large, forced-draught ovens at 30 °C after spreading on shallow trays, and hand sorting to remove roots and large stones. The dried soil is then usually crushed by heavy steel rollers in rotating cylindrical 2-mm sieves. The stones are discarded, after weighing with any stones removed earlier if stone content is required. The sieved soil would then also be weighed. Sub-samples, usually of 50–100 g, may then be taken either by careful cone-and-quartering, or with a shute splitter or spinning riffler.[7] Sub-samples are further ground, either by hand or automatically with an agate pestle and mortar or with a vibratory ball mill. Throughout care must be taken to avoid contamination with potential determinant elements.

## Rocks and Geological Samples

Sediment samples are generally treated in much the same way as soil samples, although an anti-oxidizing buffer may be added in the field to anoxic sediments if speciation studies are to be conducted, and such samples are then clearly to be kept moist.[9,10] Rocks may be broken up with a suitable crusher prior to grinding and sub-sampling. As with soils, sub-samples of *ca.* 10 g or more of coarsely ground rock will usually be further ground to allow smaller (1 g or less) sub-samples to be taken for analysis without introducing excessive sub-sampling error. The ground samples may be stored in glass vials or polythene bags, and oven dried immediately prior to analysis.

# 3  Sample Dissolution Techniques

## Waters

Acidified stored water samples are generally analysed directly, or after filtration through a 0.2 or 0.4 $\mu$ membrane filter. Particulates on filters are dissolved in a suitable digestion acid mixture, generally following dry ashing to destroy the membrane (see 'Soils, Sediments, and Rocks', below).

It should be pointed out that few elements are present in most natural waters at concentrations where flame spectroscopic techniques are directly applicable. Those that are include calcium, magnesium, sodium, potassium, and, in some samples and if conditions are very carefully optimized, manganese, iron, and aluminium. Zinc, and sometimes cadmium, may be determined directly by AFS. Mercury and hydride-forming elements may be determined if cold vapour and hydride generation sample introduction techniques are employed, as discussed in

[9] M.S. Frant and J.W. Ross, *Tappi*, 1970, **53**, 1753.
[10] M.S. Cresser, *Lab. Pract.*, 1978, 639.

Chapter 6. In all other instances, pre-concentration techniques such as solvent extraction must be employed,[11] as discussed briefly in Chapter 3, section 1.

## Plant Materials

The procedure used to dissolve plant tissue samples depends to a large extent upon the range of elements which is to be determined. In the author's laboratory, a procedure based upon digestion of 100–200 mg of ground plant material with concentrated sulfuric acid (5 ml) and then a 4% solution of 62% perchloric acid in concentrated sulfuric acid is used regularly with great success. Upto 40 samples are digested at a time in an electrically heated dry block heater, and the digestion is complete in less than 2 hours. The digests are suitable for the determination of nitrogen and phosphorus (by automated colorimetry), sodium and potassium (by AES or AAS), and calcium, magnesium, iron, manganese, and zinc by flame AAS.[12,13] If 200 mg of plant material is digested and the digest is diluted to 50 ml, copper may also be determined by flame AAS, but the precision is poor. Other authors prefer digestion with sulfuric acid plus hydrogen peroxide, which allows a similar range of elements to be determined.[14]

It is very important to run at least duplicate blanks with each batch of digestions, especially for trace element determinations. This allows correction to be made for contamination from apparatus and/or reagents. The loss of acid is small during the digestion if electrical heating is used, so that it is not essential to use digestion blanks for preparation of standards for flame spectrometric analysis. The matrix matching (see Chapter 3) is adequate if appropriate amounts of sulfuric acid and of the 4% perchloric acid in sulfuric acid solution are added to all standards.

For trace metals other than those described above, it is necessary to use larger samples of plant material. Often digestion with nitric acid, followed by perchloric acid is used to digest upto 2 g of plant tissue, with final dilution to 25 ml. This allows determination of elements such as nickel by flame AAS. However, great care is necessary when using perchloric acid because of the explosion risk. A suitable, wash-down fume hood should be available, or at least an efficient fume scrubbing system.

Some analysts prefer to use dry ashing techniques, ashing up to 20 g sub-samples in silica basins.[7] Care is then needed to make sure that losses onto siliceous ash residue or by adsorption onto the basin, which are serious problems for copper and cobalt especially, are circumvented. This may be done by extracting the residue at least twice with dilute nitric acid under infrared lamps. If the residue is finally diluted to *ca.* 25 ml with dilute nitric acid, numerous trace elements may be determined. For some elements, however, preconcentration and sensitivity enhancement by solvent extraction may still be necessary. Details of

[11] M.S. Cresser, 'Solvent Extraction in Flame Spectroscopic Analysis', Butterworths, London, 1978.
[12] T. Batey, M.S. Cresser, and I.R. Willett, *Anal. Chim. Acta*, 1974, **69**, 484.
[13] M.S. Cresser and J.W. Parsons, *Anal. Chim. Acta*, 1979, **109**, 431.
[14] S.E. Allen, H.M. Grimshaw, J.A. Parkinson, and C. Quarmby, 'Chemical Analysis of Ecological Materials', ed. S.E. Allen, Blackwell Scientific Publications, Oxford, 1974.

suitable procedures may be found in the author's book on solvent extraction in analytical flame spectrometry.[11] In such cases, it is important to put blanks through the entire ashing, digestion, and extraction procedure.

## Soils, Sediments, and Rocks

Samples in this category may be brought into solution either by fusion in a platinum crucible with a suitable fluxing agent such as sodium carbonate or lithium borate, or by digestion with an acid mixture containing hydrofluoric acid to dissolve silicates. For total element analysis, it is necessary to destroy the organic matter of soils and sediments first, usually by predigestion with nitric acid or by dry ashing.

Fusion with sodium carbonate provides a matrix with a high dissolved solids content which is not especially suitable for flame spectroscopy. The high sodium concentration may cause a stray light problem if AES is used on some instruments, and of course precludes the determination of sodium. It may result in rapid burner clogging if aluminium, silicon, and titanium are determined in very fuel-rich nitrous oxide–acetylene flames. On some instruments it becomes necessary to flick off the red hot carbon-rich deposits which occur on the burner head slot every few samples. This may be done with a suitable flat metal spatula, but great care is needed to avoid burns or melting the spatula. Sodium carbonate does not attack all minerals. For example, it has no effect upon chromite, which is better dissolved by fusion with potassium hydrogen sulfate.[15] However, unlike open digestions involving hydrofluoric acid, it does allow silicon determination by flame AAS (although the precision is not particularly good because of the poor detection limit).

Samples may be fused with lithium metaborate in a platinum crucible in a muffle furnace at 900 °C in about 30 minutes or over a Meker burner.[16] The melt may be poured directly into dilute nitric acid. In his excellent text, 'A Handbook of Silicate Rock Analysis', Potts has reviewed the effectiveness of lithium metaborate for dissolution of a wide range of minerals.[16] Those that are inefficiently attacked include bastnaesite, chalcocite, chalcopyrite, galena, ilmenite, monazite, pyrite, wolframite, zircon, chromite, sphalerite, cassiterite, and bunsenite. Thus lithium metaborate seems no better than sodium carbonate in this respect, though it does appear to be a superior matrix for flame spectrometric analysis, especially as quite low fluxing agent-to-rock ratios may be employed. Potts[16] also discusses briefly a range of other reagents suitable for specific fusion applications.

Acid digestions have the advantage that they provide sample solutions with lower dissolved solids contents than fusion methods, and allow removal of silicon as the fluoride by volatilization, thus avoiding the risk of chemical interference from large amounts of silicon. Care is needed when hydrofluoric acid is used with perchloric acid, on both safety and analytical grounds. Hydrofluoric acid causes severe burns if it comes into contact with skin, and any point of

[15] M.S. Cresser and R. Hargitt, *Anal. Chim. Acta*, 1976, **82**, 203.
[16] P.J. Potts, 'A Handbook of Silicate Rock Analysis', Blackie and Son, Glasgow, 1987.

contact must be washed immediately with copious amounts of water prior to treatment with a monosodium glutamate gel and seeking professional medical aid. The safety risks associated with perchloric acid have already been mentioned. When excess acid is being evaporated, the residue should not be heated to total dryness, or insoluble fluorides may form.[16]

Some laboratories employ operationally defined procedures to extract 'total elements' from soils, such as a one hour reflux with a mixture of boiling nitric and hydrochloric acids. Such an approach may be adequate, for example, to study the build up of elements such as zinc, cadmium, copper, lead, and nickel in sludge-treated soils. However, operationally defined procedures are much more often used to extract the portions of elements present in soils and sediments in a labile or 'plant-available' form. For example, solutions of EDTA or CDTA may be used to extract copper, zinc, manganese, and iron from soils,[17] or hydroxylamine hydrochloride may be used to extract easily reducible manganese or manganese oxide-bound trace elements.[6]

Sometimes these operationally defined procedures have a sound theoretical basis. For example, it is quite reasonable to suppose that leaching with magnesium nitrate solution will displace zinc from cation exchange sites in soils, or leaching with ammonium acetate will displace exchangeable calcium, magnesium, sodium, and potassium. Flame spectrometry, especially flame AAS, is widely used for the analysis of such extracts.

## Atmospheric Particulates

Unless particulates are being collected in a grossly polluted atmosphere, large volumes of air must be filtered to give sufficient amounts of material for analysis by flame spectrometry. The filters are then ashed and/or digested prior to analysis. Unused filters are usually used for blank preparation. Pio and Hall[18] have published a useful concise review of atmospheric particulate sampling, and a more detailed account has been written by Spurny.[19]

# 4 Slurry Atomization

Because of the difficulty and inconvenience of sample dissolution for soil and geological samples prior to analysis by atomic spectrometry, a number of authors have investigated the possibility of direct nebulization of slurry samples into flames or plasmas.[20,21] However Ebdon *et al.*[22] reported problems

[17] M. Cresser, in 'Atomic Absorption Spectrometry Theory, Design, and Applications', ed. S.J. Haswell, Elsevier, Amsterdam, 1991, p. 515.
[18] C.A. Pio and A. Hall, in 'Instrumental Analysis of Pollutants', ed. C.N. Hewitt, Elsevier Applied Science, London, 1991, p. 1.
[19] K.R. Spurny, in 'Physical and Chemical Characterization of Individual Airborne Particles', ed. K.R. Spurny, Ellis Horwood, Chichester, 1986, p. 40.
[20] M.S. Cresser, J. Armstrong, J. Dean, M.H. Ramsey, and M. Cave, *J. Anal. At. Spectrom.*, 1991, 6, 1R.
[21] M. Jerrow, I. Marr, and M. Cresser, *Anal. Proc.*, 1992, **29**, 45,
[22] L. Ebdon, M.E. Foulkes, and S. Hill, *J. Anal. At. Spectrom.*, 1990, **5**, 67.

associated with differences in both transport rates and atomization rates of refractory materials between samples and standards, even in plasmas. Nevertheless Jerrow and colleagues[21] were able to determine barium in soils and sediments by flame AAS in a nitrous oxide–acetylene flame, using a standard additions technique. The technique worked because the process of slurry preparation generated fresh cation exchange sites, and the barium, although added in solution, was adsorbed onto solid surfaces, thus effectively creating true slurry standards by solution additions.

To date, slurry nebulization has not found widespread or routine application in environmental analysis by flame spectrometry. In the author's experience the time saving is small or even non-existent, because the use of standard additions procedures is time consuming, and samples often have to be processed one at a time when slurry atomization is to be used.

# 5 Speciation

The term 'speciation' is used to describe any analytical procedure in which the amounts of an element in discrete chemical forms are determined, as opposed to the total amount of an element in the sample. For example, it may be of interest to determine the amounts of $Cr^{3+}$ and $CrO_4^{2-}$ in environmental samples, if the two ionic species have different toxicities, rather than the total amount of chromium. In this instance ion exchange may be used to separate the cationic and anionic species.[23] Sometimes all the species to be determined may be either cationic or anionic, as in the case of the determination of $Fe^{2+}$ and $Fe^{3+}$ or $S^{2-}$, $SO_3^{2-}$, and $SO_4^{2-}$. Sometimes inorganic and organically bound forms may be determined, as in the case of arsenic species in natural waters.[24] Where flame spectrometric determination has been employed in speciation studies of this type, selected key references may be found in the element-by-element section of Chapter 6.

In soil research, the term 'speciation' is often applied to operationally defined fractionation of heavy metals into five or more components.[25] Typically, water soluble, exchangeable, organically bound (which includes what is in biomass), amorphous oxide bound, crystalline oxide bound, and residual fractions are measured.[26] Sometimes residual fractions are further subdivided according to particle size distributions to give amounts in sand, silt, and clay fractions. Similar fractionation procedures are often applied to aquatic sediments.[27] In arid regions, often the calcium carbonate bound fractions of heavy metals are also measured.[28] Because of the constraints of detection limits, generally only cadmium, copper, iron, manganese, and zinc are usually monitored by flame spectrometry in such heavy metal speciation studies.[28]

From the spectroscopist's point of view, the main difference between solutions

[23] M.S. Cresser and R. Hargitt, *Anal. Chim. Acta*, 1976, **81**, 196.
[24] R.E. Sturgeon, K.W.M. Siu, S.N. Willie, and S. Berman, *Analyst (London)*, 1989, **114**, 1393.
[25] P. Federer and H. Sticher, *Chimia*, 1991, **45**, 228.
[26] L.M. Shuman, *Soil Sci.*, 1985, **140**, 11.
[27] A. Tessier, P.G.C. Campbell, and M. Bisson, *Anal. Chem.*, 1979, **51**, 844.
[28] E. El-Sayad, M.S. Cresser, M. Abd. El-Gawad, and E.A. Khater, *Microchem. J.*, 1988, **38**, 307.

obtained in fractionation studies and those for total element analysis tends to be the lower concentrations often found in the former. Thus very careful optimization and matrix matching of sample and standard solutions are essential, together with clean and careful working to minimize contamination risk.

# CHAPTER 6

# *Some Useful Accessories*

In Chapter 2, it was pointed out that one of the factors limiting detection power in flame spectrometry is transport efficiency of sample aerosol through the spray chamber. This parameter typically has a value of only 5–6% in conventional flame spectrometers. Over the past two to three decades, considerable effort has gone into improving this situation. A major advance for mercury determination came with the introduction of cold vapour sample introduction into the atomizer, exploiting the fact that this element exerts an appreciable vapour pressure of free mercury atoms even at room temperature. This was soon followed by the realization that several other elements could be converted quite readily into species such as hydrides that were gaseous at room temperature. Such gases could be transported to an atomizer with much greater efficiency than solution aerosols. These two techniques, which have dramatically improved the sensitivity of the determinations of the elements concerned, are considered briefly in sections 1 and 2. An alternative, and potentially more widely applicable approach, is to accept the limited transport efficiency, but to try to increase the residence time of atoms in the flame cell, for example by using the atom trapping techniques described in section 3.

Apart from limitations imposed by aerosol transport inefficiency, conventional analytical flame spectrometry suffers from three other significant limitations. These are the sample requirement (generally 1–2 ml per determination), the time required to add releasing agents or ionization buffers, and the limited linear and useful working ranges of calibration graphs. Sections 4–7 of this chapter consider how the effects of these restrictions may be minimized.

## 1  Cold Vapour Mercury Determination

If mercury ions in solution are reduced to the free element, and a current of air or inert gas is passed through the solution, mercury vapour, which is monatomic, will be swept out of the solution into the gas phase. This provides a very sensitive basis for the determination of this toxic element.[1] The apparatus required is illustrated in Figure 1. The flame is replaced by a glass tube atom cell with silica end windows in atomic absorption. Usually, for convenience, the atom cell is clamped to the top of a conventional AAS burner head. If atomic fluorescence is

---

[1] W.R. Hatch and W.L. Ott, *Anal. Chem.*, 1968, **40**, 2085.

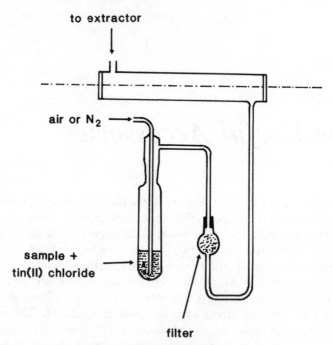

to extractor

air or N$_2$ →

sample +
tin(II) chloride →

filter

**Figure 1** *Apparatus for cold vapour mercury determination by AAS*

used for detection of the mercury atoms, usually argon is used as a carrier gas stream, and the atomic vapour is swept out of an open, circular tube immediately adjacent to a suitable source such as a small mercury vapour lamp.

Whether AAS or AFS is used, the apparatus is capable of giving detection limits in the ng ml$^{-1}$ range. Sensitivity may be even further enhanced, if necessary, by trapping the mercury released from a large sample volume as an amalgam on fine gold wire, and then liberating the mercury rapidly by heating the gold. The transient signals may be followed using a chart recorder or a suitably triggered integrator.

While the sensitivity enhancement is highly beneficial, care is needed with the technique to avoid interferences.[2] Ions which complex mercury, such as sulfide[3] and halides, may interfere with the rate of reduction to Hg. Interference may also be encountered from any volatile organic compounds in the sample which happen to absorb at the wavelength used (253.6 nm). Invariably a water trap is built into the apparatus to avoid problems caused by condensation in the atom cell.[2]

## 2 Hydride Generation Techniques

The success achieved in cold vapour mercury determinations prompted a number of investigators to examine the scope for conversion of other elements to

[2] M. Blankley, A. Henson, and K.C. Thompson, in 'Atomic Absorption Spectrometry: Theory, Design and Applications', ed. S.J.Haswell, Elsevier, Amsterdam, 1991, p. 79.
[3] L.A. Nelson, *Anal. Chem.*, 1979, **51**, 2289.

gaseous forms for sample introduction, but to simple molecular gases rather than to atomic vapours directly. The elements towards the right hand side of the periodic table are those which tend to form covalent compounds. Of these compounds, the hydrides are the most useful in the present context, because they are volatile and may be broken down readily to atoms in most cases. These hydrides are listed below, those that have been used in flame spectroscopic analysis being printed in bold type:

| Group 4 | Group 5 | Group 6 | Group 7 |
|---------|---------|---------|---------|
| $CH_4$ | $NH_3$ | $H_2O$ | HF |
| $SiH_4$ | $PH_3$ | $H_2S$ | HCl |
| **$GeH_4$** | **$AsH_3$** | **$H_2Se$** | HBr |
| **$SnH_4$** | **$SbH_3$** | **$H_2Te$** | HI |
| **$PbH_4$** | **$BiH_3$** | | |

Several of these elements, namely carbon, nitrogen, oxygen, phosphorus, sulfur, and the halogens, are not normally determined routinely by conventional AAS because their sensitive resonance lines are in the vacuum UV. Nor are these elements or arsenic, selenium, or tellurium determined by flame atomic emission techniques, because of the high excitation energies required. However, sulfur may be determined by cool flame emission techniques using a hydrogen-entrained air flame, as discussed in Chapter 2, section 12. Conversion to the gas phase dramatically reduces the interferences usually associated with cool flame emission techniques. Theoretically phosphorus may be determined in the same way, via cool flame HPO emission, but quantitative conversion of phosphate to $PH_3$ is not straightforward under analytically useful conditions.

Apart from sulfur, the elements in the table whose hydrides are printed in bold type may all be determined with excellent sensitivity by flame AAS or flame AFS techniques.[4] The hydride is generated in acidified sample solution, most commonly by the addition of sodium borohydride, $NaBH_4$, and swept into the atomizer cell with a stream of inert gas such as argon, helium, or nitrogen. The atomizer cell is most commonly a flame- or electrically-heated quartz tube, as illustrated in Figure 2. As for cold vapour mercury determinations, transient signals are obtained. Peak height may be measured on a chart recorder, or integrated absorbance or fluorescence response may be measured, depending upon the system being used.

Sodium borohydride is a strong reducing agent, and is capable of reducing a number of transition metals to the free elements, which may interfere in the hydride generation technique, especially when present in substantial excess. Interference may occur by a number of mechanisms, including consumption of the reducing agent, formation of metals which react with the hydride, adsorption of hydrogen, and disturbance of hydride transfer to the gas phase. In a sense, the

[4] T. Nakahara, in 'Sample Introduction in Atomic Spectroscopy', ed. J. Sneddon, Elsevier, Amsterdam, 1990, p. 255.

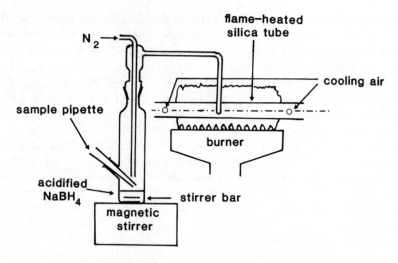

**Figure 2**  *Apparatus for the determination of elements by hydride generation using a flame-heated cell and AAS (not to scale)*

nature of the interference mechanism is of academic interest only, since whatever the cause, steps must be taken to prevent interference. Techniques which have been used to this end include separation by solvent extraction, co-precipitation, or ion exchange, and addition of masking reagents. A very useful summary can be found in a review chapter on hydride generation techniques by Taketoshi Nakahara.[4]

A surprising addition has recently been made to the list of elements which may be usefully determined by vapour generation techniques, namely cadmium.[5] Sodium tetraethylborate was used to produce a volatile cadmium species, with citrate being used to mask interference from nickel and copper. Using an argon-diluted hydrogen diffusion flame as atomizer, the detection limit by AFS was 20 ng l$^{-1}$.

## 3  Atom Trapping Techniques

Although hydride generation and cold vapour mercury determinations improve the sensitivity of the determination of a dozen or so elements by flame spectrometry, from the mid 1970s there was a widely perceived need to improve the sensitivity of the determination of other elements by flame AAS. The limitation imposed by poor transport efficiency was compounded by the short residence time of atoms in the flame. Through the early 1970s, analysts in a number of laboratories had, with limited success at least for easily atomized elements like cadmium, lead, silver, and zinc, attempted to overcome the latter problem by passing flames horizontally through long, circular quartz tubes. This

[5]  L. Ebdon, P. Goodall, S.J. Hill, P.B. Stockwell, and K.C. Thompson, *J. Anal. At. Spectrom.*, 1993, **8**, 723.

extended both residence time and absorption path length, with great benefits in terms of selectivity. However, these were achieved at the expense of a marked deterioration in selectivity as a consequence largely of atomic recombinations in the extended atom cells.[6]

Watling[7,8] made a significant breakthrough when he found that significant increases in residence time could be obtained simply by positioning a horizontal cylindrical quartz tube with a pair of parallel slots (one at the top and the other at the bottom) a short distance above the burner head supporting an air–acetylene flame. The tube was slightly longer than the flame, and the flame was channelled through the slots. The increased residence time improved sensitivity for a selection of metallic and non-metallic elements approximately three-fold.[9] This system, known as the 'Slotted Tube Atom Trap', is now commercially available.[10]

West and co-workers[11] achieved far greater sensitivity enhancement by trapping readily atomized elements, which had been nebulized conventionally for periods of up to a few minutes, onto the outer surface of a small bore, water-cooled quartz tube. If the water was then drained rapidly, the tube temperature quickly rose and the element was atomized off the surface. For cadmium in calcium chloride extracts of soil, for example, a detection limit of 4 ng (g soil)$^{-1}$ was reported. A water-cooled double silica tube atom trap similarly has been very successfully employed for the determination of cadmium and lead in natural waters by flame AAS.[12] Some further examples of applications of atom trapping are included in Chapter 7.

# 4 Sampling Cups, Boats, and Related Techniques

From the late 1960s onwards, a number of research groups around the world began to investigate alternatives to pneumatic nebulization for sample introduction, in an attempt to overcome transport efficiency limitations. The most successful approaches were those which involved heating small, discrete liquid samples, and sometimes even solid samples, directly on a metal filament, boat, or cup which could be positioned reproducibly into a flame. However, since the temperature of the metal would be lower than that of the flame itself, the techniques were confined to the determination of relatively easily atomized elements such as arsenic, bismuth, cadmium, copper, mercury, lead, selenium, silver, tellurium, thallium, and zinc.

Khan and colleagues,[13] for example, used a 50 mm long tantalum boat which

[6] G.F. Kirkbright and M. Sargent, 'Atomic Absorption and Fluorescence Spectroscopy', Academic Press, London, 1974.
[7] R.J. Watling, *Anal. Chim. Acta*, 1977, **94**, 181.
[8] R.J. Watling, *Anal. Chim. Acta*, 1978, **97**, 395.
[9] M.R. North, in 'Atomic Absorption Spectrometry: Theory, Design, and Applications', ed. S.J. Haswell, Elsevier, Amsterdam, 1991, p. 275.
[10] S.J. Haswell, in 'Atomic Absorption Spectrometry: Theory, Design, and Applications', ed. S.J. Haswell, Elsevier, Amsterdam, 1991, p. 21.
[11] S.M. Fraser, A.M. Ure, M.C. Mitchell, and T.S. West, *J. Anal. At. Spectrom.*, 1986, **1**, 19.
[12] A.A. Brown, D.J. Roberts, and K.V. Kahokola, *J. Anal. At. Spectrom.*, 1987, **2**, 201.
[13] H.L. Kahn, G.E. Peterson, and J.E. Schallis, *At. Abs. Newslett.*, 1968, **7**, 35.

was heated in an air–acetylene flame burning on a triple-slot burner head. Before being inserted into the flame centre to achieve atomization, the boat was pre-heated at the flame edge to volatilize the solvent and, if necessary, ash any organic matrix components. The transient atomic absorption signals were recorded over a period of around one second.

As might be anticipated, the reduction in flame temperature has a deleterious effect upon the incidence and extent of matrix interferences when such boat techniques are used. As a consequence, precise matrix matching is necessary for accurate results, or the standard additions method may be employed.[6] If the user is in any doubt as to whether matrix matching alone is sufficient, the adequacy of this approach may be confirmed by the analysis of certified reference materials and/or by applying the standard additions technique as well to a selection of samples to make sure that both techniques give the same results. For bismuth, cadmium, lead, silver, and thallium, detection limits by AAS are a few ng ml$^{-1}$ or better.[6] For arsenic, selenium, and tellurium they tend to lie in the range 10–30 ng ml$^{-1}$, depending upon the source used.

One of the biggest problems associated with the use of such flame-heated devices is the constraint imposed by the need for chemical inertness to matrix components, even at flame temperature, and the need to use a metal with a high melting point for the construction of the atomizer. These constraints perhaps constitute one of the major reasons for the relatively rare routine use of such techniques today compared with electrothermal atomization.

Small metal filaments have less of an adverse effect upon effective flame temperature than larger metal boats, and may produce sharp peak signals as a consequence of rapid heating. White[14] introduced solution samples into flames using small platinum wire loops. Reproducible positioning of the loop in the flame is crucial. Precision was improved by allowing the atomized cadmium or lead vapour to pass up through a hole in the bottom of a flame-heated nickel tube positioned immediately above the loop. The atoms then diffused towards the end of the tube. As little as 0.1 ng of cadmium and 1 ng of lead could be detected by this approach.

In 1970, Delves[15] described the use of 10 mm diameter nickel metal foil micro-crucibles for the atomization of lead in blood samples, after a partial pre-oxidation with hydrogen peroxide at 140 °C. The technique, which became widely known as the 'Delves Cup Technique', was extensively used for more than a decade in many laboratories around the world, and was also applied to environmental analyses such as the determination of lead in water. A flame-heated nickel tube was again used to overcome the reproducibility problems otherwise caused by the variability in the construction of individual disposable cups.

Often a major advantage of the cups and similar devices described in this section is their remarkably small sample requirement (0.1 to 0.2 mg or less). Children and factory workers alike are generally far happier parting with this amount of blood than with half a litre! Although sample availability is not that often a major problem in environmental analysis, some studies, for example of

[14] R.A. White, *Brit. Non-Ferrous Metals Res. Assoc.*, 1968, Research Report No. A1707.
[15] H.T. Delves, *Analyst (London)*, 1970, **95**, 431.

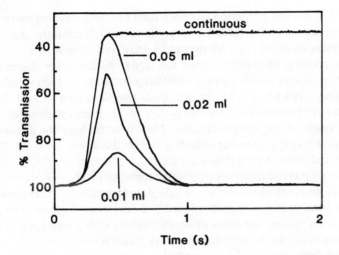

**Figure 3** *Effect of discrete sample volume aspirated and of continuous nebulization on the photomultiplier signals in AAS for a 1 mg l⁻¹ magnesium solution*

the heterogeneity of heavy metal distributions in leaves, are often only really viable if techniques with such modest sample demands are available.

# 5  Discrete Sample Nebulization Techniques

Conventional pneumatic nebulizers typically consume sample solution at the rate of *ca.* 5–8 ml min⁻¹. Thus generally, when flame spectrometry is used on a routine basis, 2–5 ml of sample solution is used per determination. However it is possible to employ much smaller volumes of sample solution.[16] Figure 3, for example shows typical atomic absorption signals for the nebulization of 0.01, 0.02, and 0.05 ml of a 1 mg l⁻¹ standard solution, as recorded on a storage oscilloscope, compared with the signal from continuous nebulization. It is clear that only about 0.04 ml of solution is required to obtain the maximum absorbance signal.

Many authors have exploited this observation in routine analysis where sample availability is limited.[17] There is some loss in precision, but this may be acceptable in many applications. There may well be other reasons why aspiration of small, discrete samples is deemed desirable. For strong acid digests, for example, damage to the nebulizer may be limited by minimizing the nebulization period. If the sample has a very high total dissolved solids content, nebulizer and/or burner head blockage may be minimized, the latter being especially important when a fuel-rich nitrous oxide–acetylene flame is being used. It may be necessary to aspirate solvents with unsuitable combustion characteristics if continuously nebulized, such as chloroform and other halogenated organic

[16] M.S. Cresser, *Anal. Chim. Acta*, 1975, **80**, 170.
[17] M.S. Cresser, *Prog. Anal. At. Spectrosc.*, 1981, **4**, 219.

solvents.[18] If solvent extraction has been used for multi-element extraction at a very high aqueous phase volume to organic phase volume ratio, the volume of organic phase available per determination may well be limited.[17]

It is interesting in Figure 3 that the signal duration for discrete sample nebulization of very small samples is virtually constant, and appreciably longer than the time taken to aspirate, say, 0.01 ml of sample. This is because the aerosol generated over perhaps only 100 ms is thoroughly mixed with all the fuel and oxidant present in the spray chamber. Thus it is the time for a complete gas change in the spray chamber which governs the lower limit on the signal duration, and not the period of aspiration.

In the early days of flame spectrometry, some very elaborate accessories were designed to give reproducible discrete sample nebulization.[16,17] However, as the technique became more widely employed, the devices used became progressively simpler, often taking the form of small funnels with a capillary bore outlet connected directly to the nebulizer capillary.[17] Even this is not really necessary, because all that is required is a small (1–2 ml capacity) beaker with a conical depression in the bottom. Conventional AutoAnalyser sample cups work very well. The end of the flexible nebulizer aspiration tube is simply dipped into the droplet of solution in the cone. This is especially useful if, for example, such sample cups have been used for evaporative pre-concentration of water samples in a vacuum desiccator.[19]

Before automated background correction became widely available in flame AAS, occasionally a limitation to detectability using discrete sample nebulization was the change in flame background sometimes caused by the introduction of water, or especially of some organic solvents, to the flame. When this problem occurred, it could have been eliminated for some determinations by using flame AFS as a preferred method of determination. However, it is largely solved if a reliable background correction system is available in AAS. In flame AES, the problem of changing background remains on most conventional instruments, however.

# 6   Devices for Addition of Releasing Agents, *etc.*

Generally in analytical flame spectrometry, the pneumatic nebulizer functions as a sample pump, as well as to produce aerosol. Most nebulizers pump sample solution at a rate somewhere between 4 and 8 ml min$^{-1}$. If the nebulizer is connected, via a suitable capillary 'T' piece, to two pieces of aspiration capillary rather than just one, one capillary may be used to aspirate sample and the other a releasing agent, ionization buffer, or even simply a diluent.[20,21] Turbulent mixing occurs at the 'T'.

Care is needed to make sure that the relative flow rates in the two arms of the 'T' are as desired. Equal flow rates will be obtained only if both uptake tubes are

[18] M.S. Cresser, 'Solvent Extraction in Flame Spectroscopic Analysis', Butterworths, London, 1978.
[19] F.A. Robertson, A.C. Edwards, and M.S. Cresser, *Analyst (London)*, 1984, **109**, 1265.
[20] I. Rubeska, M. Miksovsky, and M. Huka, *At. Abs. Newslett.*, 1975, **14**, 28.
[21] M.S. Cresser and A.C. Edwards, *Spectrochim Acta, Part B*, 1984, **39**, 609.

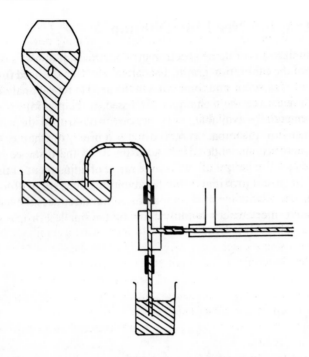

**Figure 4** *System for using a 'T' piece to continuously introduce a releasing agent*

of identical length and internal diameters, the surfaces of the sample and diluent solutions are at the same level, and the two solutions have identical viscosities.[21] Selection of different tube lengths and internal diameters allows the mixing ratio to be regulated over quite a wide range. If large diameter tubing is used, and sample and added reagent positions are not carefully maintained, there may be a marked loss in precision.[19] Figure 4 illustrates the system regularly used in the author's laboratory.

A more expensive alternative is to use standard AutoAnalyser type systems, based on multichannel peristaltic pumps, to pump samples and reagents and/or diluents at the desired rates to give automatic mixing at the desired ratio. Flame photometric detectors have been used for many years with AutoAnalysers, especially in clinical laboratories. Curiously, in the past, this approach has less often been routinely used in environmental analytical laboratories employing flame spectrometry, perhaps because an attractive feature of flame spectrometry is the speed of response when used conventionally. Over the past few years, however, there has been an increasing tendency towards fully automated, unattended operation of flame spectrometers. This undoubtedly reflects, at least in part, the improvements in safety features in modern instruments, which often incorporate a comprehensive selection of fail-safe devices. It also reflects the impact of microprocessor control systems, which have greatly facilitated automation of periodic recalibration.

# 7  Meeting the Need for Dilution

One of the limitations of flame spectrometric methods of analysis is the limited linear range of the calibration graph. This often results in the need for dilution of large numbers of samples, and sometimes in the need to repeat analyses. Dilution may be achieved using the techniques discussed in the previous section, or by using a commercially available, microprocessor-controlled dispenser–diluter. However, the use of the impact cup, as discussed briefly in Chapter 4, section 2, may be strongly recommended.[22,23] Adjustment of the distance between the impact cup and the nebulizer allows a wide range of adjustment of the transport efficiency.[23] In closest proximity, this can avoid the need for a 20-fold dilution, for example. The advantage of the impactor is that it reduces the incidence and extent of matrix interferences, by only allowing the smallest droplets to pass on into the flame.[22]

[22] M.S. Cresser, *Analyst (London)*, 1979, **104**, 792.
[23] C.E. O'Grady, I.L. Marr, and M.S. Cresser, *J. Anal. At. Spectrom.*, 1986, **1**, 51.

# CHAPTER 7
# Choice of Technique for Commonly Determined Elements

## 1 Introduction

Generally a new flame spectrometer arrives with a fairly full set of instructions on how to set the instrument up and the key parameters to use for each element that may be determined. The latter may be in a hard-copy 'cook-book' of instructions, or stored on a computer disk for rapid availability of information. Top-of-the-range instruments may even set the instrumental conditions automatically to those specified. In theory, then, there should be no need for this chapter at all. However, in practice, such manufacturers' guides often tacitly make simplifying assumptions about the sorts of samples to be analysed, and rarely tell you what to do if the instrument can't meet your needs directly. The purpose of this chapter, then, is to provide a useful guide to what can and cannot be achieved by flame spectrometric methods for each of the commonly determined elements of environmental interest. It is also intended to provide cautionary advice whenever such advice is necessary.

## 2 Aluminium

Aluminium is often determined by flame AAS, and sometimes by flame AES, which may give a marginally better detection limit at 396.2 nm than AAS. The element forms a stable bond to oxygen, and cannot be determined in an air–acetylene flame, so a nitrous oxide–acetylene flame is essential. At the temperature of this flame, however, ionization is significant, and potassium at a final concentration of $3–5 \, \text{mg} \, \text{l}^{-1}$ should be added to all samples and standards as an ionization buffer. This is especially important at low determinant concentrations. Fuel flow is critical, a fuel-rich flame with a red-feather zone of at least 5 mm being needed. Optimize the signal-to-noise ratio using a solution containing about $25 \, \text{mg} \, \text{Al} \, \text{l}^{-1}$. The optimum flow is slightly different in AAS and AES. Because of the intense emission from the flame, a narrow slit and a moderately high (e.g. 7–10 mA) lamp current usually gives the best performance. The best working range is ca. $5–100 \, \text{mg} \, \text{l}^{-1}$, although with very careful optimization the range $0.5–10 \, \text{mg} \, \text{l}^{-1}$ may be used at 396.2 or 309.3 nm. There is rarely any need to use a less sensitive wavelength.

Flame AAS is useful for the determination of total aluminium in soils and rocks, and extractable (*e.g.* with 0.1M potassium chloride) aluminium in soils. In the latter instance, the precision is better for acidic mineral soils, since concentrations in extracts from agricultural soils at near neutral pH may be very low. The extractants used for measuring amorphous aluminium in soils tend to cause severe burner clogging problems, especially in the fuel-rich flames used. It may well be necessary to flick surface deposits off from the burner head every five to ten samples, using a flat ended metal spatula. However, **this must be done rapidly and with great care**, both to avoid burns and to avoid melting the spatula. Aspirate deionized water between samples, not a reagent blank, to minimize clogging. Discrete nebulization may also be effective in this context.

There is rarely enough aluminium present in natural water samples to allow direct determination by flame AAS or AES, except when the element is organically complexed in waters draining from very acid organic suface soils, for example during storm events. Concentrations may also be measurable in waters draining mineral soils which have been severely influenced by acid deposition. Aluminium may be pre-concentrated by solvent extraction.[1] However the advantages of using an organic solvent like 4-methylpentan-2-one (MIBK) are generally less clearcut in nitrous oxide–acetylene flames than in air–acetylene, so a high water:organic solvent extraction ratio must be employed to gain any real benefit.

A low digest volume-to-sample size ratio, *e.g.* 0.5 g in 25 ml, must be used if flame spectrometry is to be employed for the determination of the aluminium content of plant materials.

# 3   Arsenic

Arsenic has its main resonance lines low in the UV, and these high energy transitions make flame AES of little use for the determination of this element. However, it may be determined by either AAS or AFS. The line at 193.7 nm generally yields slightly better sensitivity than that at 197.2 nm, although the detection limits at these two wavelengths may be similar. The nitrous oxide–acetylene flame is more transparent than air–acetylene at these low wavelengths. If available, an electrodeless discharge lamp source will generally give a superior detection limit to a hollow cathode lamp source, but since this is still unlikely to be better than *ca.* 0.1 mg l$^{-1}$, the fact is of academic interest in most environmental applications. Either the element must be preconcentrated, for example by solvent extraction,[1] or hydride generation must be employed.

Hydride generation may result in a detection limit for arsenic of around 0.8 ng ml$^{-1}$ by AAS under optimized conditions.[2] Over recent years, this impressive detection power has resulted in the development of automated, flow-injection-based hydride generation systems for the determination of arsenic in plants[3] and soil

[1] M.S. Cresser, 'Solvent Extraction in Flame Spectroscopic Analysis', Butterworths, London, 1978.
[2] M.S. Cresser and I.L. Marr, in 'Instrumental Analysis of Pollutants', ed. C.N. Hewitt, Elsevier Applied Science, London, 1991, p. 99.
[3] V. Arenas, M. Stoeppler, and H. Mueller, *Colloq, Atomspektrom. Spurenanal.*, 1989, **5**, 483.

extracts.[4] Hydride generation may also be used in speciation studies for arsenic forms in waters.[5] Many authors still seem to prefer AAS with electrothermal atomization to the use of hydride generation plus flame AAS for arsenic quantification, however. Partly this is a consequence of the complexity of interference problems when hydride generation is used.[6] Hershey and Keliher[7] have recommended ion exchange procedures for reduction in interferences in the measurement of arsenic and selenium by hydride generation AAS.

# 4 Barium

Barium forms a thermally stable oxide and is therefore only atomized to a useful extent in the nitrous oxide–acetylene flame. It may then be determined by either AAS or AES at 553.6 nm, AES generally giving a significantly better detection limit, typically by one order of magnitude. This might be expected from the low excitation energy corresponding to 553.6 nm. Addition of an ionization buffer, such as potassium at 5 g l$^{-1}$, is essential to suppress ionization. In both AAS and AES, it is crucial to carefully optimize fuel flow. Molecular band spectra of calcium species may pose problems when calcium is present at a large excess (see Chapter 3, section 3), as is generally the case in environmental samples. The problem may be minimized by adding an extra collimator to the instrument, to confine the field of view to the flame centre, since the unwanted molecular absorption and emission are greater at the flame edges.

Barium is present at very low concentrations in most environmental samples. Thus, in spite of the availability of a detection limit of only a few ng ml$^{-1}$ by flame AES, the element is rarely determined by flame methods; AAS with electrothermal atomization or ICP-AES is more commonly used. A notable exception is in the determination of the element in barium-rich geological deposits.[8] Another exception is in the analysis of formation waters from offshore oil wells.[9] However, in this matrix, inter-element interferences are encountered from alkali and alkali-earth elements. These could be effectively eliminated by the addition of 5 g l$^{-1}$ magnesium and 3 g l$^{-1}$ sodium as a modifier.[9]

# 5 Boron

Because of the great strength of the B—O bond, even in a fuel-rich nitrous oxide–acetylene flame, the atomization efficiency of boron is so low that its determination in almost all environmental materials is not possible by flame spectrometric methods. The author favours a solution spectrofluorimetric

[4] B. Weltz and M. Schubert-Jacobs, *At. Spectrosc.*, 1991, **12**, 91.
[5] J.T. Elteren, N.G. Haselager, H.A. Das, C.L. De Ligny, and J. Agterdenbos, *Anal. Chim. Acta*, 1991, **252**, 89.
[6] L.H.J. Lajunen, 'Spectrochemical Analysis by Atomic Absorption and Emission', The Royal Society of Chemistry, Cambridge, 1992.
[7] J.W. Hershey and P.N. Keliher, *Spectrochim. Acta, Part B*, 1989, **44**, 329.
[8] T. Yu. Dudnikova, R.S. Balakireva, A.E. Tavrin, and V.F. Toropova, *Zavod. Lab.*, 1988, **54**(6), 29.
[9] M. Jerrow, I.L. Marr, and M.S. Cresser, *Anal. Proc.*, 1991, **28**, 40

determination,[10] although AAS with electrothermal atomization is also useful provided a suitable matrix modifier is employed.[11]

# 6   Cadmium

With the main resonance line for cadmium at 228.8 nm, it is hardly surprising that this element is not determined usefully by flame AES. However cadmium is a very easily atomized element, and the determination by flame AAS is sensitive, with detection limits sometimes as low as 1 ng ml$^{-1}$ often being cited for the air–acetylene flame.[1] Determination by flame AFS may result in detection limits two orders of magnitude lower than this, if a suitable excitation source is available.[12] The determinations in acetylene flames are virtually free from chemical interference. Because of the ease of atomization, the element may be readily determined using atom-trapping techniques or boat or cup techniques, as discussed in Chapter 6. Recently a cold vapour sample introduction technique has also been suggested for cadmium determination.[13,14]

In spite of the sensitivity of the determination, because of the low concentrations of cadmium in most environmental samples, the element is still often preconcentrated. For example, discrete nebulization flame AAS has been used to measure foliar cadmium after extraction of the APDC complex into chloroform.[15] Cobalt was extracted at the same time. Many other solvent extraction procedures have been described.[1] Alternatively resins such as a chelating polydithiocarbamate resin have been employed to concentrate cadmium prior to determination.[16] Extractions onto solid phase materials for preconcentration may be made more convenient by automation, for example using flow injection methodology.[17]

High concentrations of cadmium are rare in environmental samples, so it is unusual to need to avoid dilution. This is perhaps just as well, because the second most sensitive wavelength for cadmium determination by flame AAS gives a sensitivity almost three orders of magnitude poorer than the main resonance line.

# 7   Calcium

Calcium has exceptionally simple atomic absorption and emission spectra, and the 422.7 nm line is used for its determination by AAS, AES, and, very rarely, AFS. The next most sensitive absorption wavelength gives some 200 times poorer sensitivity, and is of no practical interest. The element is moderately well atomized in a fuel-rich air–acetylene flame, although the determination in this

[10] I.L. Marr and M.S. Cresser, 'Environmental Chemical Analysis', Blackie and Son, Glasgow, 1983.
[11] Y.-Q. Jiang, J.-Y. Yao, and B.I. Huang, *Fenxi Huaxue*, 1989, **17**, 456.
[12] I. Rubeska, V. Svoboda, and V. Sychra, 'Atomic Fluorescence Spectroscopy', Van Nostrand Reinhold, London, 1975.
[13] L. Ebdon, P. Goodall, S.J. Hill, P.B. Stockwell, and K.C. Thompson, *J. Anal. At. Spectrom.*, 1993, **8**, 723.
[14] A. Dulivo and Y.-W. Chen, *J. Anal. At. Spectrom.*, 1989, **4**, 319.
[15] B. Weltz, S. Xu, and M. Sperling, *Appl. Spectrosc.*, 1991, **45**, 1433.
[16] M.C. Yebra-Biurrun, A. Bermejo-Barrera, and M.P. Bermejo-Barrera, *Analyst (London)*, 1991, **116**, 1033.
[17] Z.-I. Fang and B. Weltz, *J. Anal. At. Spectrom.*, 1989, **4**, 543.

flame is prone to chemical interference, especially from aluminium, phosphate, and silicate. A releasing agent, such as lanthanum at 5 mg ml$^{-1}$, is therefore essential in environmental analysis. Interferences are reduced high up in a fuel lean flame, but such conditions are not conducive to optimal sensitivity. Matrix reagents, especially phosphoric and sulfuric acids, have a marked effect upon both sensitivity and optimization of parameters such as flame height or fuel flow. Careful matrix matching is therefore invariably necessary when determining calcium.

Some analysts prefer to conduct calcium determinations in a nitrous oxide–acetylene flame to minimize the risk of interferences, and this is a sound practice. However, the element has a low ionization potential, so that an ionization buffer such as 5 mg ml$^{-1}$ potassium must then be added. The AES determination in this flame is very sensitive, and gives a lower detection limit than flame AAS. However flame AAS is sufficiently sensitive to meet the needs of most environmental applications. Flame AFS is really only of academic interest for calcium determination.

# 8 Chromium

Chromium is atomized to a reasonable extent in a fuel-rich air–acetylene flame, but the determination in this flame is highly susceptible to interference effects from a range of elements likely to be present in significant amounts in environmental samples. Moreover, chromium in different oxidation states gives a different response in this flame, and for chromium(VI) the response tends to vary with solution pH.[18] Therefore the determination is best carried out in a moderately fuel-rich nitrous oxide–acetylene flame, with careful optimization of the fuel flow for the matrix concerned. In flame AAS, the most sensitive resonance line at 357.9 nm is the one invariably employed. If necessary, Cr(III) and Cr(VI) may be measured individually using an exchange resin to separate the cationic and anionic species.[19] The speciation step may be coupled with a pre-concentration step, and the entire process automated using flow injection methodology.[20]

The detection limit by conventional flame AAS is generally around 10 ng ml$^{-1}$, the precise value depending upon burner design, among other factors. For chromium, flame AFS offers no advantages, but flame AES sometimes gives slightly better detection limits at 425.4 nm than AAS at 357.9 nm. Thus pre-concentration is often necessary before applying flame spectrometric determination of chromium to environmental samples.[1] For example, extraction of the 1,5-diphenylcarbazide has been used in water analysis.[21] However, because of the low concentrations often present, many analysts favour furnace AAS over flame techniques for this determination, to avoid the need for a time-consuming pre-concentration step.

[18] M.S. Cresser and R. Hargitt, *Talanta*, 1976, **23**, 153.
[19] M.S. Cresser and R. Hargitt, *Anal. Chim. Acta*, 1976, **81**, 196.
[20] M. Sperling, S. Xu, and B. Weltz, *Anal. Chem.*, 1992, **64**, 3101.
[21] M. Jaffar, S. Pervaiz, and M. Ahmad, *Pak. J. Sci. Indust. Res.*, 1990, **33**, 465.

# 9    Cobalt

Of the three flame techniques covered in this monograph, AAS is the only one used to any great extent for cobalt determination, the other techniques generally giving poorer detection limits. Usually a wavelength of 240.7 nm is employed, along with a lean air–acetylene flame. Even under carefully optimized conditions, the detection limit is only about 10 ng ml$^{-1}$, so that it is important to avoid any unnecessary dilution of samples. Sulfate depresses cobalt absorbance except at very low determinant concentrations, so, as usual, careful matrix matching is advisable, but generally few interferences have been reported. For most environmental samples, pre-concentration techniques such as solvent extraction are essential to achieve adequate sensitivity.[1] For example, the diethyl-dithiocarbamate complex has been used to pre-concentrate cobalt from water samples.[22]

# 10    Copper

Copper is invariably determined by AAS in a lean air–acetylene flame, using the main resonance line at 324.7 nm. The detection limit is generally around 10 ng ml$^{-1}$, which is marginally better than that generally achievable by flame AFS, and comparable to that reported for AES using a carefully optimized nitrous oxide–acetylene flame.[2] Provided samples are not excessively diluted, this value is adequate for many practical applications in environmental analysis, such as the measurement of plant copper concentration or EDTA- or DTPA-extractable copper in soils. Interferences are rare, and unlikely to be a problem from concomitant elements present in most environmental samples, but matrix matching is still advisable. The sensitivity is inadequate for the direct determination of copper in natural water samples, for which a suitable preconcentration technique must be employed.[1,23,24]

# 11    Gallium

The detection limit for gallium determination at 287.4 nm in an air–acetylene flame is only about 70 ng ml$^{-1}$, and that by flame AFS is not much better, and sometimes even worse.[1] The detection limit by flame AES at 403.3 nm is appreciably better, especially if a nitrous oxide–acetylene flame is used. This reflects the low excitation energy. These values are too low to make the direct determinations useful in environmental applications, and therefore solvent extraction is often used for pre-concentration.[1] One method often used is the extraction of the anionic keto–chloro complex from strong hydrochloric acid solution (*e.g.* 5.5M) into 4-methylpentan-2-one.[25,26] Co-extraction of iron may

[22] F. Sugimoto, Y. Maeda, and T. Azumi, *Kenkyu Hokoku – Himeji Kogyo Daigaku*, 1988, **41A**, 91.
[23] S. Cai, *Lihua Jianyan, Huaxue Fence*, 1992, **28**, 113.
[24] S. Wu and W. Fang, *Fenxi Huaxue*, 1991, **19**, 286.
[25] M.S. Cresser and J. Torrent-Castellet, *Talanta*, 1972, **19**, 1478.
[26] E.N. Pollock, *At. Abs. Newslett.*, 1971, **10**, 77.

be prevented by prior reduction to Fe(II) with an excess of titanium(III) sulfate.[25] Such an approach has been used for the determination of gallium in limonite by AAS.[26] Flame spectrometry is only really useful, even with solvent extraction, for samples with a moderately high gallium content; for other samples, AAS with electrothermal atomization or ICP-AES are more often used.

# 12 Indium

Indium has not proved to be an element of great interest in most environmental samples, in which it is usually present at very low concentrations. The flame AAS determination in a lean air–acetylene flame at 303.9 nm has a detection limit of around 50 ng ml$^{-1}$, and flame AFS is not much better.[1] Flame AES in a nitrous oxide–acetylene flame gives a much lower detection limit at 451.1 nm, of around 2 ng ml$^{-1}$. However the element has a low ionization potential, and addition of potassium at 5 mg ml$^{-1}$ as an ionization buffer is therefore advised. Sensitivity may be enhanced by solvent extraction pre-concentration using a high extraction ratio.[1] Even when pre-concentrated from geological samples by extraction into 4-methylpentan-2-one from 6M hydrochloric acid solution, ICP-AES may be the preferred method of analysis.[27]

# 13 Iron

Because of its relatively high abundance in the earth's crust, iron is often present in a wide range of environmental samples at concentrations readily determinable by flame spectrometric procedures, even without pre-concentration techniques.[1] The detection limit by AAS at a wavelength of 248.3 nm, using a carefully optimized, lean air–acetylene flame, is around 10 ng ml$^{-1}$. Flame AFS is very rarely used, because it gives no improvement with readily available sources. The element may be determined by AES at 372.0 nm using a nitrous oxide–acetylene flame, although there is no advantage in using this technique.

Silicate, nickel, and cobalt tend to interfere in the air–acetylene flame, although nickel and cobalt are rarely present in sufficient excess to cause a problem. Silicate interference may be eliminated at modest excesses by the use of lanthanum as a releasing agent or by using a nitrous oxide–acetylene flame. Very careful optimization is sometimes necessary, for example in the analysis of freshwaters, when concentrations are very low. It is important to use a narrow spectral bandpass and to make sure that the correct line is being used, because the hollow cathode lamp emission spectrum of iron is extremely complex. If you have any doubts about monochromator calibration, check the sensitivity at adjacent lines!

For some iron determinations used in soil science, extracts may contain high dissolved solids contents or organic extractants which may cause a change in sensitivity. Careful matrix matching should always be used with such extractants.

[27] Z. Tang, Z. Jin, F. Liang, and X. Zhou, *Yankuang Ceshi*, 1991, **10**, 100.

# 14   Magnesium

Magnesium is one of the elements most commonly determined by flame spectrometry, nearly always by flame AAS. The determination in a lean air–acetylene flame at 285.2 nm is remarkably sensitive by this technique, with a detection limit of around 0.1 ng ml$^{-1}$ under carefully optimized conditions. The normal working range is only around 100–500 ng ml$^{-1}$, curvature of the calibration graph typically becoming severe at around 1500 ng ml$^{-1}$. At higher concentrations, either samples must be diluted, or an impact cup may be used (see Chapter 4, Figure 6), or the alternative line at 202.6 nm may be used (useful over the range 5–20 mg l$^{-1}$).

There are a number of interferences in magnesium determination in the air–acetylene flame, the best known being silicate, phosphate, and aluminium, so a releasing agent, *e.g.* lanthanum at a final dilution of 5 mg ml$^{-1}$, must always be used.

At concentrations above 2 mg l$^{-1}$, sulfate also starts to interfere on many instruments, although no such interference may be encountered at lower concentrations.[28]

The determination of magnesium is so sensitive that there is rarely any reason for attempting to determine the element by AFS or AES. Nevertheless, the element may be determined with good sensitivity at 285.2 nm by AES using a nitrous oxide–acetylene flame, although potassium should then be added as an ionization buffer, especially at very low magnesium concentrations.

# 15   Manganese

Manganese may be determined with good sensitivity by flame AAS or AFS, the former technique being very widely used in environmental analysis. The detection limit at 279.5 nm by AAS in a lean air–acetylene flame is about 10 ng ml$^{-1}$, which is quite adequate for most environmental analyses. A narrow spectral bandpass should be used, and care taken to make sure that the 279.5 nm line is being used rather than one of the adjacent lines at 279.8 or 280.1 nm. The detection limit at 403.1 nm by flame AES using a nitrous oxide–acetylene flame is usually slightly better, at around 5 ng ml$^{-1}$, than that obtained by flame AAS.

In natural water samples, concentrations of manganese may often be at or below the detection limit of flame AAS, and solvent extraction or some other technique must be used for pre-concentration.[29–32]

# 16   Mercury

As explained in Chapter 6, section 1, the determination of mercury at 253.7 nm using conventional nebulization and flame AAS or flame AFS is not particularly

[28] M.S. Cresser and D.A. MacLeod, *Analyst (London)*, 1976, **101**, 86.
[29] M.N. Ghandi and S.M. Khopkar, *Anal. Sci.*, 1992, **8**, 233.
[30] K. Dagmar, *Int. J. Environ. Anal. Chem.*, 1991, **45**, 159.
[31] S.A. Abbasi, *Rev. Roum. Chim.*, 1991, **36**, 901.
[32] M.C. Yebra-Biurrun, M.C. Garcia-Dopazo, A. Bermejo-Barrera, and M.P. Bermejo-Barrera, *Talanta*, 1992, **39**, 671.

sensitive. The detection limit in AAS using an air–acetylene flame is only about 500 ng ml$^{-1}$, for example.[2] This is far too high a concentration to be of any practical value in almost all environmental analyses. By cold vapour techniques, detection limits of around 1 ng ml$^{-1}$ are generally attainable by AAS and around 0.1 ng ml$^{-1}$ by AFS. Even lower detection limits are attainable if a gold amalgamator is used for pre-concentration.[33] Alternatively, in water analysis, the element may be pre-concentrated by solvent extraction prior to determination by cold vapour AAS or AFS.[34]

# 17 Molybdenum

The abundance of molybdenum in the earth's crust is low. Thus the detection limit for the determination of the element by AAS at 313.3 nm in a slightly reducing nitrous oxide–acetylene flame, at around 30–50 ng ml$^{-1}$, is inadequate for its routine determination in most environmental samples without pre-concentration.[1]

The element is extractable from strong hydrochloric acid solutions into 4-methylpentan-2-one. This approach may be applied to the analysis of plant material, if the ash is extracted with the strong hydrochloric acid required.[35] Kim *et al.*[36] masked iron(III) by reduction to iron(II) with tin(II) chloride before extracting molybdenum as its thiocyanate complex with Aliquat 336 into chloroform. The latter was evaporated, and the residue extracted with 4-methylpentan-2-one prior to determination of molybdenum by AAS. The procedure was applied to soils, sediments, and natural waters. In fertilizer analysis, the thiocyanate complex of molybdenum has been extracted, after reduction of iron with tin(II) chloride, into 3-methylbutan-1-ol, and the latter extract analysed directly.[37] In another thiocyanate-based procedure, total molybdenum from soils and geological materials was extracted into 4-methyl-pentan-2-one.[38]

# 18 Nickel

Nickel is determined more often, and with better sensitivity, by flame AAS than by flame AFS or AES techniques, even when a nitrous oxide–acetylene flame is employed in AES. The AAS detection limit at 232.0 nm under carefully optimized conditions in an oxidizing air–acetylene flame is about 10 ng ml$^{-1}$, which is adequate for those environmental applications where a low solution-to-sample weight ratio may be used. For example, if 1 g of plant material is digested with a mixture of nitric acid plus perchloric acid, and the mixture diluted only to 10 ml, nickel may be determined directly by flame AAS. Similarly the

[33] L. Liang and N.S. Bloom, *J. Anal. At. Spectrom.*, 1993, **8**, 591.
[34] Y. Liu and S. Zhao, *Lihua Jianyan, Huaxue Fence*, 1991, **27**, 359.
[35] H.M. Nakagawa, J.R. Watterson, and F.N. Ward, *US Geol. Surv. Bull.*, 1975, No.1408, 29.
[36] C.H. Kim, P.W. Alexander, and L.E. Smythe, *Talanta*, 1976, **23**, 229.
[37] W.L. Hoover and S.C. Duren, *J. Assoc. Off. Anal. Chem.*, 1967, **50**, 1269.
[38] C.H. Kim, C.M. Owens, and L.E. Smythe, *Talanta*, 1974, **21**, 445.

author has been able to analyse *aqua regia* digests of sludge-treated soils. Nickel hollow cathode lamps give a complex spectrum in the vicinity of the main resonance line at 232.0 nm, and it is very important to use a narrow spectral band-pass and to make sure that the correct line has been selected. If in doubt, compare sensitivity with that at immediately adjacent lines. Passing too much light at adjacent wavelengths results in very pronounced curvature, and indeed some manufacturers recommend using the three-times poorer sensitivity wavelength of 341.5 nm to obtain a wider linear calibration range. There are few problematic interferences in environmental samples at low determinant concentrations.

For many samples, pre-concentration is essential, and this is commonly achieved by solvent extraction. Often the nickel tetramethylenedithiocarbamate complex is extracted at pH 2–4 into 4-methylpentan-2-one.[1] This system has been applied to soil and sediment extracts[39] and to water samples.[40–42] Kinrade and Van Loon[43] used a mixture of ammonium tetramethylenedithiocarbamate and diethylammonium diethyldithiocarbamate to extract a range of elements, including nickel, into 4-methylpentan-2-one from water samples adjusted to pH 5. New solvent extraction-based procedures are still being published regularly for environmental samples such as plant tissues and water samples.[44]

# 19 Phosphorus

Although phosphorus may be determined by flame AAS at 213.6 nm using a fuel-rich nitrous oxide–acetylene flame, the sensitivity of the determination is so poor that it is of no practical worth in the present context; nor are flame AFS or AES techniques. Solution spectrophotometric or colorimetric procedures, or higher temperature AES techniques such as ICP-AES, or sometimes ion chromatography, are much more useful. The only reason the element is included here is because of the interesting studies which have been carried out on its determination by cool flame molecular emission techniques, as discussed briefly in Chapter 2, section 12.[45] While such determinations are highly susceptible to chemical interferences, as might be expected at the low temperature used, they have application when flame photometric detectors are coupled to chromatographic separation systems.

# 20 Potassium

Potassium, because of its exceptionally low excitation potential, is still very widely determined at 766.5 nm using filter flame photometry. The low temperature flames used in flame photometry have the advantage that ionization of the

[39] T.T. Chao and R.F. Sanzolone, *J. Res. US Geol. Surv.*, 1973, **1**, 681.
[40] R.R. Brooks, B.J. Presley, and I.R. Kaplan, *Talanta*, 1967, **14**, 809.
[41] K.M. Aldous, D.G. Mitchell, and K.W. Jackson, *Anal. Chem.*, 1975, **47**, 1034.
[42] A. Heres, *Analusis*, 1972, **1**, 408.
[43] J.D. Kinrade and J.C. Van Loon, *Anal. Chem.*, 1974, **46**, 1894.
[44] M.S. Cresser, J. Armstrong, J. Cook, J.R. Dean, P. Watkins, and M. Cave, *J. Anal. At. Spectrom.*, 1994, **9**, 25R.
[45] R.M. Dagnall, K.C. Thompson, and T.S. West, *Analyst (London)*, 1968, **93**, 72.

element is generally negligible, so it is unnecessary to add an ionization buffer. This is certainly not true if the element is excited in a nitrous oxide–acetylene flame, where caesium at least 2 mg l⁻¹ must be added. However a sub-ng ml⁻¹ detection limit may then be achieved by AES, provided the photomultiplier fitted is sufficiently red-sensitive. Even in air–acetylene, potassium at low concentrations is significantly ionized unless an efficient ionization buffer is added.

An ionization buffer should also be added when the element is determined by AAS at 766.5 or 769.9 nm (the latter wavelength giving approximately three times poorer sensitivity) in an oxidizing air–acetylene flame. Absence of an ionization buffer results in concave curvature of the calibration graph at low concentrations.

# 21 Selenium

Selenium cannot be usefully determined by flame AES techniques, because of the high excitation energy associated with its low-wavelength resonance lines. However, it has been shown that it can be determined by a cool flame, molecular emission cavity analysis technique, as discussed briefly in Chapter 2, section 13.

The element may be determined at 196.0 nm by AAS, using a nitrous oxide–acetylene flame (which is more transparent than air–acetylene at this low wavelength), or by AFS in a variety of flames.[46,47] The detection limit of both techniques for selenium is around 1 mg l⁻¹, too low to be useful for environmental analyses. The element is therefore invariably determined by hydride generation techniques, coupled to AAS or AFS detection, as discussed in Chapter 6, section 2, or by furnace AAS, or occasionally by solution spectrofluorimetry using 2,3-diaminonaphthalene as a reagent. If direct flame AAS or AFS are to be used for some reason, then pre-concentration by solvent extraction is necessary.[1] However, this approach is rarely used nowadays.

# 22 Sodium

Sodium is still often determined by flame photometry, measuring the emission intensity of the doublet at around 589 nm, but care is necessary to make sure that excess calcium does not cause spectral interference (from molecular emission). This is unlikely to be a problem if AES is used, with a narrow spectral band-pass, and the intensity of emission at 589.0 nm from an air–acetylene flame is measured. However, at low determinant concentrations it is then advisable to add 2–5 mg ml⁻¹ potassium or caesium as an ionization buffer. This is even more true if a nitrous oxide–acetylene flame is used for FES, although its use is rarely justified in environmental analyses because the additional sensitivity gained is rarely necessary.

Flame AAS may also be used for sodium determination, although this offers

[46] R.F. Browner, *Analyst (London)*, 1974, **99**, 617.
[47] N. Omenetto and J.D. Winefordner, *Prog. Anal. At. Spectrosc.*, 1979, **2**, 1.

no real advantage over AES for environmental samples. Some analysts believe it is unnecessary to invest in a sodium hollow cathode lamp.

## 23  Strontium

The most sensitive flame spectrometric procedure for the determination of strontium is FES, the emission intensity at 460.7 nm being measured from a nitrous oxide–acetylene flame. A detection limit of 1 ng ml$^{-1}$ or better is generally readily attainable, although the element has a low ionization potential and addition of potassium or caesium at a final concentration of 2–5 mg ml$^{-1}$ is essential as an ionization buffer. Chemical interference from phosphate, silicate and aluminium is reduced dramatically in this flame.

Strontium may also be determined at the same wavelength by AAS, using a nitrous oxide–acetylene flame and ionization buffer to minimize the risk of interference. Although slightly poorer by AAS than by AES on most instruments, the detection limit is still as low as a few ng ml$^{-1}$.

## 24  Sulfur

Sulfur is never determined routinely in environmental samples by flame atomic spectrometric techniques, because its lines lie in the vacuum UV. However, the determinations of sulfide and sulfite or trapped sulfur dioxide using gas-phase sample introduction and cool-flame emission spectrometry have been shown to be both sensitive and selective, and capable of automation using autoanalyser methodology.[48,49] Sulfite may be converted to sulfur dioxide simply by acidification, or to hydrogen sulfide by addition of sodium borohydride.[49] Sulfide in samples is converted to hydrogen sulfide by simple acidification. The evolved gaseous sulfur compound is swept into a hydrogen diffusion flame and the blue $S_2$ emission intensity is measured. Detection limits for sulfur are 100 ng ml$^{-1}$ or better. Similar methodology is used for sulfur-specific detectors for chromatography.

## 25  Tellurium

Because of the low wavelengths of its main resonance lines, tellurium is never routinely determined by flame AES, but the element may be determined with moderate sensitivity by flame AAS or AFS techniques, at 214.3 nm in an oxidizing air–acetylene flame. The detection limit is generally around 100 ng ml$^{-1}$, which is inadequate for almost all environmental applications unless the element is pre-concentrated by a substantial factor.[1] However, hydride generation plus AFS or AAS yields much more useful detection limits. Even so, pre-concentration may be necessary if the element is to be determined in natural waters, for example by extraction as the tributylphosphate complex.[50] The

[48]  T.A. Arowolo and M.S. Cresser, *Talanta*, 1992, **39**, 1471.
[49]  R.M. Dagnall, K.C. Thompson, and T.S. West, *Analyst (London)*, 1969, **94**, 643.
[50]  Y. Zheng, J. Zexiang, X.-G. Liu, and P. Wang, *Fenxi Huaxue*, 1987, **15**, 966.

element has attracted little interest from environmental scientists, perhaps because of its very low natural concentrations.

# 26 Tin

Tin is a problem element in terms of its routine determination in environmental samples by conventional flame spectrometry. Partly this is because of the poor sensitivity (the detection limits for all three techniques are around 100 ng ml$^{-1}$ or higher), and partly it is because of the low natural abundance of the element in most environmental samples. Moreover, it is an element that is not well suited to pre-concentration by solvent extraction techniques, although recently a procedure has been described for soil acetate extracts in which tin was extracted with crown ether into chloroform after treatment of the extract with picric acid.[51]

Tin is best determined by AAS at 224.6 or 235.5 nm in a reducing nitrous oxide–acetylene flame. Hydrogen as a fuel may give improved sensitivity, but only at the expense of serious interference problems.

Hydride generation techniques are applicable to tin, and the detection limit is then improved dramatically, generally to around 1 ng ml$^{-1}$. For natural water samples, the element is still sometimes pre-concentrated prior to determination by hydride generation techniques.[52,53]

Because of its toxicity, tributyl tin is a compound which is often determined in contaminated water samples and sediments. It may be determined by cold vapour techniques using AAS or by gas chromatography with flame photometric detection, although a round-robin exercise has shown that it is difficult to attain satisfactory precision.[53] Great care is needed in terms of a clean working environment and to follow specified procedures exactly. It has been suggested that, for sea water analysis, after generation, the hydride from tributyl tin should be concentrated by temporary cryogenic trapping at $-196\,°C$.[54] This approach allowed an impressive detection limit of <4 pg ml$^{-1}$. A somewhat similar approach has been applied to the analysis of acetic acid extracts of sediments, although the tributyl tin was isolated by gas chromatographic separation after trapping and prior to the final determination stage.[55]

# 27 Titanium

Being an element that is apparently not particularly toxic, not particularly soluble, and not essential to biota, titanium has not been determined often in environmental samples. It forms a very stable bond to oxygen, so a fuel-rich nitrous oxide–acetylene flame is essential for its determination. By AAS the detection limit at 364.3 nm is generally around 100 ng ml$^{-1}$, or slightly better

[51] V.P. Poluyanov, *Khim. Sel'sk, Khoz.*, 1992, **4**, 99, cited in *J. Anal. At. Spectrom.*, 1993, **8**, 402R.
[52] S. Nakashima, *Fresenius' J. Anal. Chem.*, 1992, **343**, 614.
[53] J.W. Readman and L.D. Mee, *Mar. Environ. Res.*, 1991, **32**, 19.
[54] J.-J. Chen, Y.-C. Lou, and C.-W. Whang, *J. Chin. Chem. Soc.*, (*Taipei*), 1992, **39**, 461.
[55] P.H. Dowson, J.M. Bubb, and J.N. Lester, *Mar. Pollut. Bull.*, 1992, **24**, 492.

than this value. Similar results are generally found by flame AES at 399.9 nm. However, the determinations require addition of 2–5 mg ml$^{-1}$ potassium or caesium as an ionization buffer, and usually excess aluminium, typically around 1 mg ml$^{-1}$, is also added to prevent interference from aluminium, iron, and fluoride. If necessary, the element may be pre-concentrated by solvent extraction.

# 28  Zinc

Because it is such an easily atomized element with such a simple spectrum, zinc may be determined with excellent sensitivity by flame AAS, the detection limit at 213.9 nm generally being about 1 ng ml$^{-1}$ when an air–acetylene flame is used. This is more than adequate for many environmental applications, and interferences are rarely a problem, although as always matrix matching is advisable. If, however, a lower detection limit is deemed necessary, and a suitable source is available, AFS may yield a detection limit at least one order of magnitude lower.[12] Alternatively atom trapping techniques, as described in Chapter 6, or solvent extraction[1] may be used to further enhance sensitivity. Flame AES is not used for routine zinc determinations because of the low wavelength, and hence high excitation energy, of the main zinc resonance line.

# 29  Other Elements

The coverage of elements given above is selective rather than exhaustive. Other elements could, without doubt, be determined in environmental samples by flame spectrometry. If they are not included above, it implies that serious practical limitations probably would be encountered, and that, at least in the author's opinion, flame spectrometric methods should not be chosen except as a last resort.

# 30  Speciation and Hybrid Techniques

From time to time in this chapter, mention has been made of procedures for the determination of particular chemical forms of elements. Sometimes this is achieved by a totally manual fractionation of an element into separate and distinct chemical fractions, and the fractions are then analysed one at a time by spectrometric procedures. However, by the 1980s there was a growing tendency to automate the fractionation procedures and to incorporate the spectrometric detector 'on-line'. It is appropriate at this point to consider a few examples of this approach.

Tye *et al.*[56] used hydride generation coupled to AAS to quantify organic and inorganic arsenic species in soil pore waters, after pre-concentration on a pellicular anion-exchange column. They were able to detect down to 2 ng of arsenate, arsenite, and monomethyl arsenite and down to 1 ng of dimethyl arsonate. More recently, an argon–hydrogen–entrained air flame fitted with a

[56] C.T. Tye, S.J. Haswell, P. O'Neill, and K.C.C. Bancroft, *Anal. Chim. Acta*, 1985, **169**, 195.

slotted tube atom trap has been used to separate seven arsenic compounds after separation by high pressure liquid chromatography.[57]

Inorganic, methyl, and n-butyl tin compounds may be converted to volatile hydrides, and the latter separated on a chromatographic column prior to detection of tin by AAS.[58] In this particular study, an electrically heated absorption cell was used, although a flame-heated quartz cell could have been employed equally well. Balls[59] used on-line cryogenic trapping on a silanized glass-wool column to separate dibutyl tin and tributyl tin in sea water prior to transport to a quartz-tube atomizer for determination by AAS.

Speciation of lead in air and atmospheric particulates is still a topic of great environmental relevance. Sodium tetrahydroborate may be used to hydrogenate inorganic lead ions and alkyl-lead species.[60] As in the study by Balls outlined above, cryogenic trapping may again be used to trap temporarily the plumbane and alkyl- and tetraalkyl-lead compounds, which are then released sequentially by heating for detection by flame AAS.

It will be clear from the above examples that hydride-forming elements especially are often subjected to such speciation studies. There are, of course, good reasons for this. These are the very elements which tend to form toxic organometallic compounds, and they are elements which may be determined with excellent sensitivity. Moreover, interferences are not usually a problem following a separation process.

[57] S.H. Hansen, E.H. Larsen, G. Pritzl, and C. Cornett, *J. Anal. At. Spectrom.*, 1992, **7**, 629.
[58] O.F.X. Donard, S. Rapsomanikis, and J.H. Weber, *Anal. Chem.*, 1986, **58**, 772.
[59] P.W. Balls, *Anal. Chim. Acta.*, 1987, **197**, 309.
[60] P. Foster, M. Laffond, R. Perraud, P. Baussand, and V. Jacob, *Int. J. Environ. Anal. Chem.*, 1987, **28**, 105.

# CHAPTER 8

# How Do I Know I'm Getting the Right Answer?

## 1 Introduction

Some readers might find it disconcerting that there should be any need to ask the above question of readers, after they have ploughed diligently through seven chapters of detailed advice and guidance. While it is true that many of the limitations of flame spectrometric methods of analysis are now well documented and understood, it should never be forgotten that environmental samples naturally provide a complex matrix for analysis, and often this matrix contains large excesses of a number of concomitant elements. Interference studies are, in most instances, performed by looking at effects of one concomitant element at a time, at one or two concomitant ion concentrations and at a single determinant concentration, on a single instrument under a single set of operating conditions. It has long been realized that extrapolation from conclusions drawn about selectivity under one set of conditions to diverse conditions introduces a risk of systematic error.[1] However, most books on flame spectrometric methods are full of such extrapolations. Thus there is considerable scope for unsuspected interferences occurring unless suitable tests to confirm accuracy are performed. Even if there are no unsuspected interferences, it is still possible to get the wrong answer.

## 2 The Need for Quality Control

Over recent years, there has been increasing emphasis upon quality control in the production of analytical data. This is hardly surprising, since there can be little point in performing an analysis if the accuracy of the determination is of no consequence. In most spectrometric methods there is considerable scope for producing poor results, because the methods are virtually all secondary methods, relying upon the accuracy of standard solutions for calibration. An error in the preparation of a litre of 1 mg l$^{-1}$ stock metal solution could result in years of data with a systematic error unless adequate checks are made.

Provided the balances and volumetric apparatus used for standard preparation have been properly calibrated, carefully prepared standards should not introduce

---

[1] M.S. Cresser, *Lab. Pract.*, March 1977, 171.

significant error. If there is any doubt, for example about the purity of a reagent or about its moisture content as a result of drying, a new stock solution should be checked by gravimetric or titrimetric methods of analysis. Whenever a new stock solution is prepared, the new dilute standards prepared from it should be compared with the remains of the previous set of standards. This is especially important if long-term changes in environmental parameters are being assessed.

For both consistency over each day, and to check day-to-day reproducibility, it is very useful to analyse at regular intervals either synthetic standard solutions, prepared in an appropriate matrix, or, better still, standard or certified reference materials, which contain precisely known amounts of the elements or species of interest. Routine long-term time-plots of the results of such regularly repeated analyses are very useful in showing up errors arising as a consequence of hitherto unrealized procedural changes.

# 3 Reference Materials

Certified reference materials are now available for many environmental materials, and are often used in setting up new methods or for confirming that modifications to existing procedures are acceptable. If the results for a selection of four to five such materials covering a range of determinant and potential interferent concentrations are very close to certified values, this is strong circumstantial evidence that the method is working well.

Certified reference materials are expensive to produce and correspondingly expensive to purchase. They should therefore be used sparingly and looked after carefully to avoid contamination. Many analysts prefer to produce their own standard reference materials in bulk for quality control purposes. Care is needed to make sure that such a standard material is homogenized as well as possible, which usually involves producing a material with a very fine particle size.

There is an advantage to using both reference materials and synthetic standards for quality control monitoring. The former show up errors arising at any stage in the laboratory phase of the analytical procedure. Thus, they will throw light upon any problems arising during sample preparation stages or during the flame spectrometric determination stage. Synthetic standards highlight problems arising in the latter stage only.

# 4 Inter-laboratory Comparisons

An alternative method for performance evaluation is to participate in inter-laboratory comparisons, often known as 'round-robins'. Usually this involves sending sub-samples of a selection of appropriate samples to a number of independent laboratories, to be analysed either by a fully specified procedure and technique, or to be analysed by whatever method each laboratory choses. Obviously, these two approaches test different things. The former indicates the precision attainable using a specified procedure, and tests both the adequacy of the specification and the competence of participating laboratories. Only the latter is tested in the second approach, although such a round-robin may show up the inadequacy of the methods chosen by some laboratories.

It should be stressed that inter-laboratory comparisons are not a substitute for regular, in house quality control procedures. Their main value is in novel procedure evaluations or, especially in the present context, when several different laboratories are collaborating in an integrated environmental study.

# 5   Other Procedures for Error Detection

There may well, on occasions, be people reading this book who wish to make sure that their data is reliable, but who cannot afford either the time delay or the cash outlay needed to acquire a selection of appropriate certified reference materials. Fortunately, there are other useful approaches worthy of consideration which allow detection of spectral, chemical, or transport interferences in analytical flame spectrometry.

The easiest way to check for the absence of a spectral interference is to reanalyse the sample using one or more different wavelengths for the determination. It is highly improbable that the same 'wrong answer' will be obtained at two different wavelengths, and even less likely that the same erroneous answer will be obtained at three different wavelengths.

The extent of chemical interferences in flame spectrometry varies with flame conditions and analyte concentration. Thus it is most unlikely that the same wrong answer will be obtained at two different heights in the flame or at two different fuel-to-oxidant ratios. Indeed it has been suggested that the former of these two options may provide automated detection of chemical interferences.[2] The burner was moved up and down using a microprocessor-controlled stepper motor. Alternatively, results in air–acetylene and nitrous oxide–acetylene flames may be compared. Similarly, it is unlikely that the same wrong answer will be obtained at two different dilutions. Thus if it is thought that there might be a risk of chemical interference, the determinations on a selection of samples should be repeated under diverse conditions, either on the same or different instruments.

The commonest way of confirming the absence of physical or transport-related interferences is to see if the standard additions method gives the same results as direct nebulization of samples and standards (see Chapter 3, section 1). If it does, physical interference is most unlikely. It is possible to measure aspiration rate continuously in flame spectrometry,[3] although this is not yet a standard practice. One approach to doing this is to connect a small pressure transducer, via a capillary 'T' piece, at right angles to the aspiration tubing. The pressure drop between the internal tip of the nebulizer capillary and the 'T' joint is a function of the aspiration rate. Such a system may be used to provide early warning of nebulizer blockage.

If a collimated light source is fitted inside the spray chamber, and a photodiode detector is used to measure the amount of light scattered by aerosol droplets inside the spray chamber, it is possible to detect directly changes in aerosol

[2] A.A. Gilbert, I.L. Marr, and M.S. Cresser, *Microchem. J.*, 1991, **44**, 117.
[3] C.E. O'Grady, I.L. Marr, and M.S. Cresser, *Analyst (London)*, 1985, **110**, 431.

transport.[4] Once again, this technique is not applied routinely in environmental analysis.

The above suggestions do not constitute an exhaustive list, but should provide ideas for sufficient easily applied tests to allow the analyst to have greater confidence in the accuracy of his data. However, if time allows, it is still better, in addition, to analyse authentic or simulated reference materials.

[4] P. Gordon and M.S. Cresser, *Analyst (London)*, 1989, **114**, 1389.

# CHAPTER 9
# *Safety in Flame Spectrometry*

## 1 Introduction

Aspects of safety have been touched upon in Chapter 2, and especially in section 5 of that chapter. However, there are other aspects which need to be considered, at least briefly, even in an introductory text such as this one. Accidents could arise in analytical laboratories routinely employing flame spectrometry for a number of reasons, which include:

(i) handling heavy cylinders of gases;
(ii) handling gases which may be inflammable and/or toxic, or, in the case of nitrous oxide, increase risk of fire;
(iii) risk from handling high-temperature flames and burner heads;
(iv) risks associated with toxic fume generation; and
(v) risks associated with the use of hazardous reagents during sample preparation.

Categories i–iv are those most obviously associated with flame spectrometry, but risks under group v above should not be overlooked, although they come under more generalized good laboratory safety practice. In laboratories employing several analysts, for example, it is all to easy for labelling of sample in digest vials to be processed by flame spectrometry to consist only of sample identification numbers, even if the digests contain potentially hazardous reagents such as hydrofluoric acid or strong oxidizing acids such as perchloric acid. Adequate labelling, with risks clearly identified, is still essential, especially if samples are passed from one analyst to another at some stage. It is equally essential that facilities should be clearly identifiable, and immediately to hand, to deal with any spillages or breakages.

## 2 Handling Gases

If possible, gases should be kept in secure cages outside the laboratory building, and piped into the laboratory for routine use through appropriate and clearly identifiable metal piping. This offers a number of advantages. It means that all gas cylinders are handled at ground floor level in the open air, where accidental, sudden large-scale gas release or longer-term slow gas leaks would be less of a problem, and bulky gas cylinders do not clutter up the laboratory. Travelling with fuel cylinders in lifts is not a practice to be encouraged. In the event of fire in a building, the obvious hazards may be considerably reduced by being able to shut off gas supplies outside, especially assuming that fuel and oxidizing gas cylinders are caged separately. Also it should not be forgotten that acetylene is toxic, as well as inflammable.

If cylinders are to be brought into the laboratory, they should only be transported on purpose-built trolleys, to which they can be securely chained during transport. Acetylene cylinders must be kept more or less vertical, for reasons discussed in Chapter 2, section 5. The total number of cylinders in the laboratory at any one time must be kept to the minimum possible, especially for fuels and oxidants, and in routine use all cylinders must be securely chained or strapped to a bench or wall. They should be positioned for easy access and so that they will not block a rapid exit. It is perfectly feasible to run two flame spectrometers from a single acetylene cylinder, via a 'T' junction, although individual flash-back arrestors should be fitted in the lines to both instruments. If an acetylene cylinder is used inside the laboratory, it is especially important to check for gas leaks whenever the cylinder head is changed. Don't rely on the smell of escaping gas, which will only detect fairly major leaks. Use of a paint brush and a soap solution is more reliable, as bubbles will be clearly seen if gas is escaping.

If the laboratory is to be vacated for more than the briefest period, flames should be extinguished and cylinders inside the laboratory turned off at the main valve. Cylinders external to the working laboratory should be turned off at the main valve at the end of routine work periods, *e.g.* daily or at lunch time, *etc.*

# 3   Safety Devices and Safety Checks

The wide range of safety features and interlocks built into modern instruments has already been discussed in Chapter 2, and these will not be considered again here. However on older instruments it is still important to check that the safety diaphragm or blow-out bung is in place and intact, and that the drain contains an adequate amount of water. The author still makes his students carry out these rapid and simple checks even on well-interfaced instruments. He also routinely makes them himself. He is aware of at least one occasion when a minor flashback caused a small tear in the safety diaphragm and a subsequent serious leak, but did not trigger the safety interlock which would have prevented the flame from being relit.[1] Some manufacturers, for reasons best known to themselves, position the safety diaphragm out of sight at the rear of the spray chamber. If the diaphragm is checked by feel, take care not to burn your wrist on a previously used burner head. This is one warning instruments still do not generally provide.

It is advisable to check all gas lines, both within and external to the spectrometers, on a regular basis for leaks. Fitting fuel gas sensors which would automatically provide early warning even of very slight fuel leaks is a more reliable way of tackling this potential, although hopefully rare, problem.

[1] M.S. Cresser, *J. Anal. At. Spectrom.*, 1988, **3**, 301.

# 4 Fume Extraction

Extraction hoods serve a number of useful purposes. These include removal of carbon dioxide and steam (the commonest combustion products) and of other, less desirable combustion products, such as sulfur dioxide and other acidic gases from some sample solutions, and of whatever elements were present in the samples. They also remove soot from excessively fuel-rich flames, which otherwise can make a real mess in the laboratory. It is unwise to run instruments for extended periods without fume extraction, even if there is nothing obviously toxic in the sample solutions. This is especially true for the nitrous oxide–acetylene flame. It is well worth considering interfacing the fume extraction switch with the instrument power supply, because it is easy to forget to turn on the extractor when everything else is automated.

# 5 Instruction Sheets

Instruction sheets or booklets to be used by inexperienced operators must be clear, unambiguous, and provide safety prompts at appropriate stages. At the same time they should not be so lengthy, or have so many cross references to other sections, that they are not user friendly. It is generally better to have an adequate two or three page sheet for novices than to let them loose with a 50-page instruction book on their first 'outing'. Knowledge is best assimilated in easily digestible stages.

# Subject Index